Resilient Policies in Asian Cities

Mitsuru Tanaka • Kenshi Baba
Editors

Resilient Policies in Asian Cities

Adaptation to Climate Change and Natural Disasters

Springer

Editors
Mitsuru Tanaka
Hosei University
Machida, Tokyo, Japan

Kenshi Baba
Tokyo City University
Yokohama, Kanagawa, Japan

ISBN 978-981-13-8598-8 ISBN 978-981-13-8600-8 (eBook)
https://doi.org/10.1007/978-981-13-8600-8

This Springer imprint is published by the registered company Springer Nature Singapore Pte Ltd.
The registered company address is: 152 Beach Road, #21-01/04 Gateway East, Singapore 189721, Singapore

Preface

Japan suffered severe damage due to the Great East Japan Earthquake in March 2011, which was the largest earthquake and tsunami on record. This disaster led to a widespread focus on the prevention and minimization of damage from disasters and on keyword such as "resilience" and "resilient" from post-damage reconstruction perspective. These concepts refer to the ability of the citizens or the existence of social systems that would allow society to appropriately deal with and recover from external impacts (external forces) of severe natural disasters, such as earthquakes, volcanic eruptions, typhoons, and floods. These terms also refer to a social structure that will permit the protection of citizens' safety, lives, and health once society has obtained these abilities. Resilience could be said to be one of the most essential factors of modern society.

Indeed, after the Great East Japan Earthquake, the term "resilience" was adopted in Japan from the perspective of the ability to develop a strong resistance to, quickly recover from, and adapt to the damage from disasters. Resilience formed the basis of the keyword "national resilience," which was established in 2013 via the *Basic Law for National Resilience that Contributes to Disaster Prevention and Reduction.*

The international community has also adopted the concept of resilience as it relates to the abovementioned natural disasters, and there are more than a few discussions about information networks, energy supply, ecology/biodiversity, and climate change from a broad standpoint. For example, in 1970, Holling stated that resilience within an ecological context was "system resilience, the ability to absorb change and disturbance, and the ability to maintain the relationships between system components." Furthermore, the general public defines resilience as the strength of resistance to strong shocks and stresses and the speed of mental and physical recovery. It is thus used in the psychological and medical fields.

Regarding the concepts of "resilient" and "resilience," which have many meanings, this book will focus on the environmental/energy context and will discuss the application of resilience in regional societies and cities in particular.

In Part I, the concept of resilience will be systematically organized per previous studies, and, while discussing the meaning of resilience, the types of indices for grasping and measuring resilience will be developed and proposed, and the results of the trial implementation in large Japanese cities will be summarized.

In Part II, a case study will be conducted based upon an analysis of individual applications of the concept of resilience in Asian cities, including those of Japan, the Philippines, and Thailand, and a methodology for constructing a resilient regional and local society will be proposed.

In Part III, we will refer to the mayor's initiative to establish and spread the concept of resilience as a summary of this book, and we will summarize the results and proposals developed by this research.

Although Asian countries, including Japan, are home to many diverse regions with unique features, they also have delicate and fragile natural environments, as well as areas that are at risk of large-scale disasters. Additionally, as we entered the twenty-first century, the effects of climate change have been actualized in various locations globally. As identified in the IPCC Fifth Assessment Report, disaster risks accompanying abnormal weather are predicted to increase and accelerate. The risks associated with climate change, and large-scale disasters that are expected to bring, take many forms over a broad range. These include enormous typhoons and hurricanes, sea level rise, the intensification of storm surges along the shore, heavy rain and floods, increased frequency of landslides in inland areas, the effects of rising temperatures on agriculture and lowered food production, increased heat stress and health hazards, the spread of infectious diseases, and the progression of water shortages and desertification. Regarding the expansion of climate change risks, the creation of responsive, resilient regional communities is being pursued, and this is thought to be the primary factor in the recent major focus on "adaptation" for climate change issues.

Thus, this book reanalyzes and reorganizes the concept of resilience, proposes paths to implement resilient regional societies that can adapt to risks per said concepts, and facilitates understanding of today's concerns. We sincerely hope that this book will be used to spread practical initiatives everywhere and that as many readers as possible will be able to use this book.

In conclusion, this book is based upon research results from studies implemented via support from the Ministry of the Environment's Environment Research and Technology Development Fund (1–1304) and the Ministry of Education, Culture, Sports, Science and Technology's research fund "Social Implementation Program on Climate Change Adaptation Technology (SI-CAT)," and it summarizes the knowledge obtained by the said studies. We would like to express once again our gratitude for all of the support our research received.

Machida, Tokyo, Japan — Mitsuru Tanaka
Yokohama, Kanagawa, Japan — Kenshi Baba

Contents

Part I Perspectives

1 **A Framework and Indicators of Resilience** 3
Kenshi Baba, Yu Nagata, Shun Kawakubo, and Mitsuru Tanaka

2 **Assessment of City Resilience Using Urban Indicators in Japanese Cities** 47
Shun Kawakubo, Kenshi Baba, Mitsuru Tanaka, Shuzo Murakami, and Toshiharu Ikaga

3 **Civil Indicator: General Public's Cognitive Structure of Policies for Making Resilient Cities** 61
Kenshi Baba, Kosuke Shirai, and Mitsuru Tanaka

4 **Civil Indicator: The Resilience Index of Regional Communities to the Risks of Disasters** 85
Motoko Kosugi, Kenshi Baba, and Mitsuru Tanaka

5 **Administrative Indicator: Local Officials' Cognitive Structure of Policies for Making Resilient Cities** 105
Kenshi Baba, Kosuke Shirai, and Mitsuru Tanaka

Part II Case Studies

6 **Application of the Policy Model in Sendai: the Experiences of the Sendai Framework for Disaster Risk Reduction Action UNISDR/WCDRR Public Forum, 2015–2030** 129
Kenshi Baba, Kosuke Shirai, Yu Nagata, Shun Kawakubo, Motoko Kosugi, Mitsuru Tanaka, and Ebru A. Gencer

7 **Multi-criteria Evaluation of Local Energy Resilience** 153
Tsuneo Takeuchi

8 **Enhancing Capacities for Building Climate and Disaster-Resilient Cities in Asia: Case Study of Cebu, Philippines and Nonthaburi, Thailand** 169
Premakumara Jagath Dickella Gamaralalage, Toshizo Maeda, and Simon Gilby

9 **Strengthening Urban Resilience/Disaster Risk Management in Asian Cities** 195
Tadashi Matsumoto

10 **Knowledge Co-production Processes for Building Disaster Resilience of Communities in Coastal Areas: A Case Study of Baler, Aurora, Philippines** 205
Pedcris Miralles Orencio

Part III Recommendations

11 **"Covenant of Mayors Japan": Regional Re-creation and Global Contributions Through Climate Policy and Energy Autonomy** 231
Noriko Sugiyama

12 **Comprehensive Lessons Learned and the Next Steps** 241
Mitsuru Tanaka and Kenshi Baba

Part I
Perspectives

Chapter 1
A Framework and Indicators of Resilience

Kenshi Baba, Yu Nagata, Shun Kawakubo, and Mitsuru Tanaka

Abstract As an introduction of this book, we first examine the definitions and scopes of resilience and concepts of a resilient city using an extensive literature review. We then employ the concept and framework of assessing a resilient city and develop its indicators. Consequently, we define a resilient city as being capable of responding to multiple environmental risks, and we assume that a combination of precautionary, adaptive, and transformative measures is required according to the degree of external forces (risks or stresses). We introduce the policy model based on the assumption that the state of implementation and preparation of resilience measures is governed by three major elements: risks of external forces, vulnerabilities, and situations to be avoided. The policy model also includes three types of indicators—urban, citizen, and administrative indicators—which measure the state of each element of the policy model.

Keywords Climate change · Natural disaster · External force · Vulnerability

1.1 Introduction

Today, the term "resilience" is prevalent when discussing a sustainable society. Particularly in Japan, following the Great East Japan Earthquake, related discussions have advanced rapidly in the Advisory Committee on National Resilience (Disaster Reduction and Mitigation), by the National Resilience Promotion Office

K. Baba (✉)
Tokyo City University, Yokohama, Kanagawa, Japan
e-mail: kbaba@tcu.ac.jp

Y. Nagata
Kumamoto City Office, Kumamoto, Japan

S. Kawakubo · M. Tanaka
Hosei University, Machida, Tokyo, Japan
e-mail: kawakubo@hosei.ac.jp; mtanaka@hosei.ac.jp

M. Tanaka, K. Baba (eds.), *Resilient Policies in Asian Cities*,
https://doi.org/10.1007/978-981-13-8600-8_1

of the Cabinet Secretariat, since spring 2013. However, as indicated by the title of the Basic Act for Building National Resilience (Disaster Reduction and Mitigation), which serves as the basis of national resilience policy, and by the deliberations of the Advisory Committee (National Resilience Promotion Office, Cabinet Secretariat 2014), the most attention is paid to disaster reduction and mitigation.

Examining resilience is a global trend that is not limited to Japan alone, and there has been a variety of proposals concerning the concept of a resilient city. However, it must be noted that, in more than a few cases worldwide, the concept of a resilient city is discussed in terms of major environmental policy issues, whereas the issues are almost restricted to disaster reduction and mitigation in Japan. Although it is one of the important aspects in discussion of resilience, globally recognized resilience is also discussed in the context of environmental policy (hereinafter "environmental resilience"), especially climate change, which covers a broad range of issues such as energy, ecosystems, wastes, green urbanism, and so on.

Accordingly, in this chapter, we first examine the definitions and scopes of resilience and concepts of a resilient city by a broad range of literature review and interviews with the officials of local governments. Then, based on the findings of the review, we employ the concept and framework of assessing resilient city and develop the indicators for it. Finally, we summarize the characteristics of the concept, framework, and indicators in several guidelines of resilience.

1.2 Methodology

A literature review was conducted as follows. We first used a search engine to find articles on resilience. By using the key word "resilience," we got more than approximately 32,000 articles; we then selected approximately 60 articles that have the top impact factors in the field of environmental sciences, biodiversity, engineering, water resources, public administration, geography sociology, and urban studies. In addition to that, we collected other articles derived from the above articles and Japanese articles and books.

While arranging the concept, we also conducted interviews with the local officials (the general affairs department, the environmental policy department of the city of Kawasaki, the environmental policy department and other departments of the city of Sendai, the planning and administration department and other departments of the city of Toyota, and the environmental policy department and other departments of the city of Nagoya). We also obtained feedbacks from the officials of the local governments and those of the National Resilience Promotion Office of the Cabinet Secretariat and the Global Environment Bureau of the Ministry of the Environment of Japan at the "Resilient City Workshop" that we hosted in Hosei University on February 20, 2014.

After clarifying the concept and framework of resilience by interviews and workshops, we scrutinize various administrative plans such as comprehensive plans, environment master plans, and community disaster-prevention plans in the above-mentioned local governments to identify indicators for the assessment of resilience. At the same time, we referred to several guidelines of resilience to develop our understanding of the concepts, frameworks, and indicators. The common purpose of these guidelines is to make cities and communities more resilient, but the framework and indicators differ greatly from one another. The summary of the characteristics of each guideline on resilience will be given at the end of this chapter.

1.3 Framework of Resilience

1.3.1 Various Definitions of Resilience

First, we outline the concept of resilience based on the literature review. A large number of surveys has already been conducted in a variety of fields, such as Ishihara and Nakamaru (2007) at the psychological and individual level; Shiozaki and Kato (2012), Norris et al. (2008), Manyena (2006), and others at the disaster reduction and the local community levels; Mori (2010), Resilient Alliance (2002), and others in the field of ecology and social ecosystems; and Fujii et al. (2012) in the field of economics. Referring to these, Table 1.1 presents the definitions from a number of previous studies likely to be useful as references in considering the subjects of this book—resilient city and environmental resilience.

A frequently cited study that discussed the concept of resilience is Holling (1973), which holds that resilience is a concept expressing the properties of an ecosystem with regard to environmental changes. It has since been cited in a very large number of papers from a variety of fields. The above-mentioned Norris et al. (2008) identifies that the ability to adapt to turbulence, stress, and disaster is a point that has been stressed in many definitions of resilience and suggests that there is a consensus on the two points that resilience is better conceptualized as a process than a result and as a form of adaptability rather than stability. Next, it argues that between engineering resilience (returning systems to their previously designed states and functions after a disturbance) and ecological resilience (tolerating a variety of desirable conditions suited to the environment), the latter is better suited to human beings, communities, organizations, and society. It then offers its own definition, arguing that resilience is achieved when robustness, redundancy, and speed counteract stress factors, and that resilience is, in fact, a networked combination of adaptabilities. This definition includes some very broad-ranging aspects, formed by linking economic development, information and telecommunications, community abilities, and social capital.

Table 1.1 Examples of the concepts of resilience

Author, year of publication	Area	Definition
Holling 1973	Ecosystem	According to this definition, resilience is the property of the system; resilience determines the persistence of relationships within a system and is a measure of the ability of these systems to absorb changes of state variables, driving variables, and parameters and still persist
Adger 2000	Society	This article argues that social resilience is defined as the ability of communities to withstand external shocks to their social infrastructure
Resilient Alliance 2002	Social ecology	"Ecosystem resilience" is the capacity of an ecosystem to tolerate disturbance without collapsing into a qualitatively different state that is controlled by a different set of processes. A resilient ecosystem can withstand shocks and rebuild itself when necessary
		Resilience in social systems has the added capacity of humans to anticipate and plan for the future
		"Resilience" has three defining characteristics: The amount of change the system can undergo and still retain the same controls on function and structure; the degree to which the system is capable of self-organization; the ability to build and increase the capacity for learning and adaptation
Godschalk 2003	Urban	A resilient city is a sustainable network of physical systems and human communities. During a disaster, the physical systems must be able to survive and function under extreme stresses
UNISDR 2005	Urban	The capacity of a system, community, or society potentially exposed to hazards to adapt, by resisting or changing in order to reach and maintain an acceptable level of functioning and structure. This is determined by the degree to which the social system is capable of organizing itself to increase this capacity for learning from past disasters for better future protection and to improve risk reduction measures
Norris et al. 2008	Local community	Community resilience is a process linking a network of adaptive capacities (resources with dynamic attributes) to adaptation after a disturbance or adversity. Thus, in summary, we propose that resilience resources have three dynamic properties: robustness, redundancy, and rapidity
		Community resilience emerges from four primary sets of adaptive capacities—economic development, social capital, information and communication, and community competence—that together provide a strategy for disaster readiness

Source: Holling (1973), Adger (2000), Resilient Alliance (2002), Godschalk (2003), UNISDR (2005), Norris et al. (2008)

1.3.2 Various Definitions of Resilient City

One key international policy development regarding the resilient city is the "Making Cities Resilient Campaign" started in 2010 under the UNISDR (United Nations International Strategy for Disaster Reduction) (2005). Based on the Hyogo Framework for Action (HFA) 2005–2015, which was adopted at the second World Conference for Disaster Reduction held in Kobe in 2005 with the participation of 168 national governments, 78 regional and international agencies, and 161 NGOs, the campaign is intended to strengthen the understanding of and commitment to the mitigation of disaster risk and improvement of resilience, as well as increasing the priority of climate change policies among national and local governments. This would ensure that each national government's efforts are more likely to reduce the vulnerabilities and disaster risks that they need to address. It is expected that the cities participating in the campaign will improve their resilience through learning from each other and mutual technical assistance.

The UNISDR calls for developing disaster resilience as a factor in achieving sustainable development, and it provides an overview of key strategies and actions needed to build resilience to disasters, as part of an overall strategy to achieve sustainable development. Climate change and extreme weather events are likely to increase the city's exposure to hazards and risks. Here, risk is a function of the hazard (a cyclone, an earthquake, a flood, or a fire, for example), the exposure of people and assets to the hazard, and the conditions of vulnerability of the exposed population or assets. Outlining the specific details of the state of such disaster resilience, it is one: (1) where disasters are minimized because the population lives in homes and neighborhoods with organized services and infrastructure that adhere to sensible building codes, without informal settlements built on flood plains or steep slopes because no other land is available; (2) that has an inclusive, competent, and accountable local government that is concerned about sustainable urbanization and commits the necessary resources to develop capacities to manage and organize itself before, during, and after a natural hazard event; (3) where the local authorities and the population understand their risks and develop a shared, local information based on disaster losses, hazards, and risks, including who are exposed and who are vulnerable; (4) where people are empowered to participate, decide, and plan their city together with local authorities and value local and indigenous knowledge, capacities, and resources; (5) that has taken steps to anticipate and mitigate the impact of disasters by incorporating monitoring and early warning technologies to protect infrastructure, community assets, and individuals, including their homes and possessions, cultural heritage, and environmental and economic capital, and is able to minimize physical and social losses arising from extreme weather events, earthquakes, or other natural or human-induced hazards; and (6) that is able to respond, implement immediate recovery strategies, and quickly restore basic services to resume social, institutional, and economic activity after such an event.

Since 2010, the International Council for Local Environmental Initiatives (ICLEI) also has held an annual international conference under the title "Resilient

Cities." The ICLEI's concept of the resilient city is as follows: For a city to be sustainable, it must be resilient to disasters, climate change, and unforeseeable events, and in order to improve its resilience, it must reduce its exposure and vulnerability to risks while also increasing its resistance and robustness and preparing for emergencies (ICLEI 2012). The ICLEI, founded in 1990 at the World Congress of Local Governments for a Sustainable Future held at the United Nations, is an international network of local governments and local government associations dedicated to sustainable development. At present, its membership consists of more than 1000 local governments from 84 countries. Every year Resilient Cities welcomes more than 500 attendees who mainly take part in broad-ranging discussions concerning resilient cities prepared for the external forces of climate change and natural disasters (Otto-Zimmermann 2011, 2012).

In addition, ResilientCity.org, which is an open Internet forum operated mainly by a group of architectural and urban planning researchers and practitioners in Canada, defines a resilient city as "one that has developed capacities to help absorb future shocks and stresses to its social, economic, and technical systems and infrastructures so as to still be able to maintain essentially the same functions, structures, systems, and identity." It advocates increasing resilience from the perspectives of architectural design and urban planning, mainly to address climate change and instability in energy supplies. It advocates six principles of resilient design, including diversity (of the various systems that comprise cities) and redundancy (of infrastructure, including electrical power, fuel supply, waste water processing, and most importantly, food and potable water supply) (Resilient City.org 2013).

While the concept of resilient city has been discussed mainly in the context of cities' resilience to natural disasters, Newman et al. (2009) expand the concept to include resilience to shortages of natural resources and the effects of human activities with regard to climate change. Specifically, in their definition of a resilient city, they include that a city can substantially reduce its dependence on petroleum fuels in a way that is socially and economically acceptable and feasible. For this reason, they argue that a resilient city has built-in systems that can adapt to change, such as a diversity of transport and land use systems, and multiple sources of renewable power that will allow a city to survive shortages in fuel supplies. Furthermore, they propose the following ten principles as strategies for realizing resilient cities: (1) Set the vision, prepare an implementation strategy; (2) learn on the job; (3) target public buildings, parking, and road structures as green icons; (4) build TOD (transit-oriented development), POD (pedestrian-oriented development), and GOD (green-oriented development) together; (5) transition to resilient infrastructure step by step; (6) use prices to drive change where possible; (7) rethink rural regions with reduced oil dependence; (8) regenerate households and neighborhoods; (9) facilitate localism; and (10) use approvals to regulate for the post-oil transition.

Much of the literature (e.g., Tobin 1999) uses both the terms "sustainable city" and/or "eco city" in a similar way as "resilient city." In Japan, among the concepts such as eco city or smart city, one typical example of a specific policy is the Eco-Model City Project promoted by the Regional Revitalization Bureau of the Cabinet Secretariat of Japan. It identifies Eco-Model Cities as cities that take leadership on efforts to greatly reduce greenhouse gas emissions, as the first step toward a low-carbon society. The basic stance of these cities is one of reducing or mitigating environmental impact or external forces (risks or stresses). On the other hand, as discussed above, the concept of resilient city is mainly focused on responding to external forces. Accordingly, in this book, we consider a sustainable city to be an overarching, higher level concept, while a city that intends to mitigate external forces is referred to as eco city or smart city, in a complementary relationship with the resilient city.

1.3.3 The Risk in the Context of Environmental Policy in Japan

Looking at the context of environmental policy in Japan, risk is defined as the possibility that the use of a technology or attendant human actions or activities could have undesirable effects on human health or safety, property, or the environment (systems). Environmental risk is regarded as one of the risks as well as risk of natural disaster, risk of urban disaster, risk of food security, and others (Ikeda and Morioka 1993).

According to the (then-) Environment Agency of Japan, the term environmental risk was first used in Japan in the first Environment Master Plan, formulated in 1994. Later, "the report of the colloquium on environmental protection in the 21st century (1996)" established within the Environment Agency defined environmental risk as the possibility that the environmental burden of human activities could, under certain conditions and through processes within the environment, affect health or ecosystems.

While the environment master plan mainly considered the use of chemicals as an environmental risk factor, such factors can include any and all of the factors that could cause impediments to environmental protection, such as modifications to the environment and greenhouse gas emissions (Uchiyama 2006).

1.3.4 Definition and Scope of Resilience

Based on the findings of the above review, we assume that natural disasters and climate change, which influence each other, are among the major human-caused external forces (risks or stresses) and that these impact various facets of socio-ecosystems. Along with the premise, we define the resilient city and the scope of environmental resilience as following: resilient city is capable of responding to multiple environmental risks, taking into consideration the relationship between the external forces (risks or stresses) and socio-ecosystems.

Figure 1.1 organizes the concept to give a practical form to environmental resilience. Here, up to a certain level of risk exposure (scale of environmental changes),

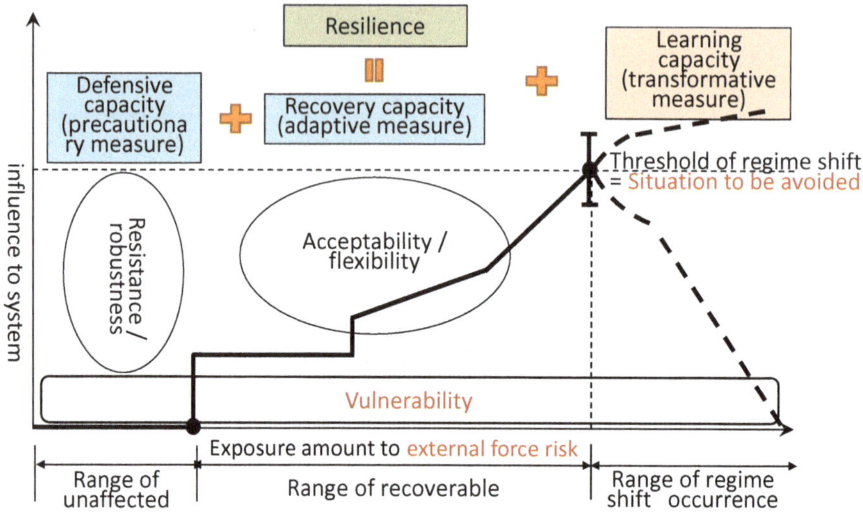

Fig. 1.1 A concept of three measures of resilient policy. (Altered from Mens et al. 2011)

urban systems will remain completely unaffected and systems will be maintained in their prior forms due to their resistance and robustness. However, when risk exposure (scale of environmental changes) exceeds one level, the impacts on urban systems will begin to appear in a discontinuous manner. Still, even at this stage, the urban systems will continue to maintain their prior forms due to their acceptability and flexibility. Measures that can be taken at this stage include precautionary measures and adaptive measures to draw out the defensive capacity and recovery capacity of the systems. Based on this understanding, there is a need to implement both these types of measures before and after such events. Furthermore, when risk exposure (scale of environmental changes) surpasses a certain threshold, a regime shift (or revolutionary phenomenon) will take place, breaking down the existing framework of urban systems. Therefore, transformative measures need to draw out learning capacity to create fundamentally new systems over a very long term. Accordingly, to make cities and communities more resilient, a combination of precautionary, adaptive, and transformative measures is required.

1.3.5 Policy Model, Status Report, and Scenario

We developed a framework that employs the terms of policy model, status report, and scenario for a resilient city, using analysis and implementation by identifying the corresponding measured assessment indicators (Fig. 1.2).

A policy model is a hypothetical flow expressing the overall process of developing policy, based on the assumption that the state of preparation and implementation of resilience measures are governed by three major elements; risks of external

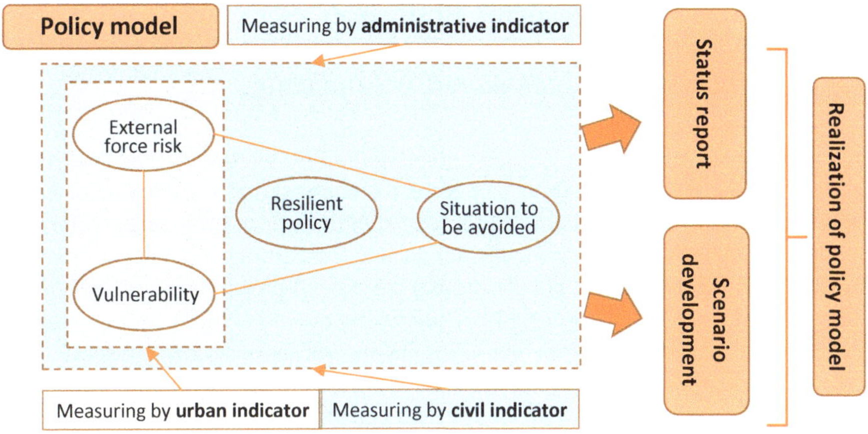

Fig. 1.2 A framework and indicators of resilient policy

forces, vulnerabilities, and situations to be avoided. It also provides the three types of indicators: urban, civil, and administrative indicators, which measure the state of each element of the policy model.

A status report is used to diagnose the resilience of each city, together with the results of measurement by the indicators. Policy scenario refers to the scenario developed in venues such as internal workshops within local government agencies and citizen conferences to implement policy, using the status report.

The details of three indicators are as following. The background of providing the three types of indicators is that the integration of expert knowledge, local knowledge, and living wisdom collected in these ways can increase environmental resilience effectively. The way of thinking about this coincides with community-based adaptation such as Allen (2006) and van Aalst et al. (2008).

- Urban indicators: These involve ascertaining and evaluation, by local government officials and experts, of resilience related to the state of factors such as the city's physical infrastructure, its economic activities, and environmental factors. They employ statistical data such as population census.
- Civil indicators: These involve ascertaining and evaluation, by stakeholders and citizens as well as experts, of resilience related to the state of the lives of citizens, including knowledge and awareness, learning and training, and social capital as well as environmental factors. They employ data from questionnaire surveys of citizens, supplemented by statistical data such as those from a public opinion poll.
- Administrative indicators: These involve local government officials and experts checking whether any relevant measures have been implemented in the past, their extent and progress, and ascertaining and evaluating whether they led to improvements in the resilience of the city. They employ data from questionnaire surveys of administrators supplemented by administrative plans and other information.

1.4 Indicators of Measuring Resilience

1.4.1 Administrative Indicators/Civil Indicators

After the above preparation and comprehensive consideration, we identified 41 indicators concerning risks from natural and social external forces anticipated in past measures, 28 indicators concerning vulnerabilities inherent to local communities and within the local government, 24 indicators concerning anticipated situations to be avoided, and 44 indicators concerning the state of preparation and implementation of resilience measures capable of addressing these. Forty-four administrative indicators consist of three types measures, that is, precautionary measures (19 indicators), adaptive measures (14 indicators), and transformative measures (11 indicators). For the civil indicators concerning acceptability of resilience measures, 44 indicators are reduced to 16 by integrating indicators similar to each other to facilitate understanding of the citizens. The administrative indicators will be described in detail in Chap. 5, and the civil indicators will be described in detail in Chap. 3 and 4 (Table 1.2).

1.4.2 Urban Indicators

Urban indicators were developed specifically for assessing resilience by the following steps: Many indicators for measuring resilience were proposed at first. The proposed indicators were then validated and carefully selected for data availability, simplicity, comparability, representativeness, and balance. Data availability is one of the most important criteria for selecting indicators, because even a theoretically sound indicator is useless if its value cannot be calculated because of the lack of data. Simplicity is also important as it helps indicator users to understand the actual condition of the target city. Comparability must be considered as well because it is difficult to understand the strengths and weaknesses of a target city in comparison with other cities if the accuracy and definition of the data vary from area to area.

Consideration of representativeness and balance is also essential when there are many candidate indicators for a single assessment factor. After developing the set of urban indicators considering the above, the urban indicators were incorporated into three comprehensive resilience indicators for assessing a city's (1) defensive capacity, which evaluates the capacity to prevent the occurrence of damage with precautionary measures; (2) recovery capacity, which evaluates the capacity to minimize damage after a disaster with adaptive measures; and (3) learning capacity, which evaluates the capacity to recover quickly from the disaster with transformative measures. Finally, the Tokyo metropolitan area and other selected cities were assessed by using the developed indicators. Over 30 indicators were initially proposed, and 18 were carefully selected according to the aforementioned criteria. Table 1.3 lists

Table 1.2 Administrative/civil indicators to assess a resilient city

41 Indicators of risk perception		28 Iindicators of vulnerability	24 Indicators of situation to be avoided
Noise, vibration	Animal damage	Presence of lowland or land at 0 m elevation	Direct damage to human life
Soil pollution	Harmful insects	Presence of steep terrain	Long-term physical or mental health damage
Land subsidence	Increase in invasive species	Presence of rapidly flowing rivers	Cut-off of food or lifeline supply
Air pollution, odor	Decrease in/loss of biodiversity	Presence of coastlines subject to erosion	Long-term worsening of food conditions
Water contamination	Forest depletion	Presence of active volcanoes	Long-term worsening of water resources
Drought, depletion of water sources (water resources)	Decrease in food and agriculture production capacity	Presence of earthquake zones or fault lines	Worsening of living environment
Acid rain	Infectious diseases, viruses	Lack of usable water resources	Loss of ease of living/comfort
Torrential rainfall	Factory explosions, accidents	Presence of rare or endangered species	Full or partial collapse of buildings, or building damage
Heat waves, fierce heat	Chemical pollution, accidents	Single-crop farming	Cut-off or interruption of transportation and telecommunications functions
Cold waves, blizzards	Transportation accidents	Vulnerable infrastructure	Deteriorating (aged) urban infrastructure
Ocean pollution	Energy-infrastructure accidents	Presence of industrial zones	Cut-off of energy supply
Rising sea levels	Accidents at nuclear power facilities	Concentration of housing in areas prone to disaster	Long-term instability in energy supply
High tides	Accidents in information and telecommunications infrastructure	High number or density of wooden homes	Cut-off of financial service functions
Red tides	Increasing greenhouse gas emissions	High number of vacant homes	Suspension of industrial activities or supply chains
Tornadoes, strong winds	Rapid population increases	Lack of open space	Long-term decline in economic activity
Typhoons	Population decreases, low birth rates	Lack of evacuation sites	Suspension of administrative activities
Mudslides, landslides	Aging of population	Lack of medical services	Long-term decrease in the level of administrative services

(continued)

Table 1.2 (continued)

41 Indicators of risk perception		28 Iindicators of vulnerability	24 Indicators of situation to be avoided
Flooding	Poverty, disparities in living standards	High percentage of low-income earners, poor employment conditions	Temporary loss of order in society
Earthquakes	Financial bankruptcy	High number of single-person households	Chronic worsening of public safety
Tsunamis	War, disputes, terrorism	High percentage of elderly population, depopulation	Degeneration of local culture/traditions
Volcanic eruptions		Weak ties within communities	Sudden, localized worsening of natural environment
		Frequent relocation of residents into and out of the community (low retention)	Loss of shores, rural land, green land, etc.
		Lack of activities by citizens' groups, nonprofits, etc.	Long-term negative impact on ecosystems
		Weak ties between government and citizens	Increase in global warming
		Presence of political conflict	
		Lack of resources for drafting and promoting policies	
		Conservativeness of organizations in government agencies	
		Lack of tax revenue	

19 Indicators of precautionary measures	14 Indicators of adaptive measures	11 Indicators of transformative measures
Development of buildings and infrastructure compliant with current standards	Strengthening of lifeline backup functions	Resettlement from high-risk areas
Development of various disaster prevention functions and facilities compliant with current standards	Rapid provision of shelters and temporary housing, etc.	Construction regulations and land use control in high-risk areas
Promotion of renewable energy	Strengthening of firefighting and emergency medical services	Building and infrastructure improvements exceeding current standards

(continued)

Table 1.2 (continued)

19 Indicators of precautionary measures	14 Indicators of adaptive measures	11 Indicators of transformative measures
Promotion of energy conservation	Support for those who require it	Condensation of urban functions (creation of compact cities)
Nature conservation and promotion of forestation	Prompt restoration support for transportation, communication, and energy supply functions	Transfer of urban functions
Preventive measures for health maintenance	Expanding methods of gathering and providing damage information	Development of and support for next-generation telecommunications infrastructures
Support during government shutdowns	Operation of government data backup systems	Development of and support for next-generation energy infrastructures
Disaster prevention training and public awareness	Prompt transitions to emergency structures by government organizations	Establishment of and support for local energy companies
Stimulation of self-assistance, cooperative assistance, and community functions	Activity to maintain public order	Deregulation via the special ward system, etc.
Dissemination of risk information	Enhancement of public self-assistance and cooperative assistance support capacity during disasters	Implementation of and support for next-generation technical research and development
Expansion and revision of disaster hazard areas, etc.	Engagement of recovery specialists and advisors	Promotion of green infrastructures
Strengthening of various monitoring functions	Prompt establishment of support reception structures	
Accumulation of government data and coordination with policies	Various measures to prevent expansion of secondary damage and injuries	
Collection and application of scientific prediction information	Protection of traditional cultural assets	
Strengthening through penalties and various disaster prevention regulations		
Public recognition and commendation of exemplary disaster prevention initiatives		
Engagement of disaster prevention specialists and advisors		
Formation of disaster agreements		
Promotion of preservation of tradition and culture		

Table 1.3 Assessment items and corresponding urban indicators

No.	Assessment item	Urban indicator	Unit
01	Prevention of fire breakout in density area	Number of fires/ population of densely-inhabited districts	Number/ 1,000,000 people
02	Prevention of deterioration of industry production	Value of industry production/ amount of water consumption	1,000,000 yen/ m^3/day
03	Prevention of evacuation route cutoff	Total length of roads with width less than 5.5 m/ total length of all type of roads	%
04	Prevention of accidents in an emergency situation	Number of traffic accidents/ total population	Number/ 1,000,000 people
05	Prevention of delay in evacuation	Number of people requiring long-term care or support/ total population	%
06	Prevention of collapse of houses	Number of houses built before 1980/ total number of houses	%
07	Prevention of housing vacancies	Number of housing vacancies/ total number of houses	%
08	Prevention of uncomfortable room temperature	Number of houses with double-sash windows/ total number of houses	%
09	Adequacy of area for evacuation	Number of schools that can be used for evacuation/ total population	Number/ 1,000,000 people
10	Adequacy of medical professionals	Number of doctors/ total population	Number/ 1,000,000 people
11	Adequacy of medical facilities	Number of hospital beds/ total population	Number/ 1,000,000 people
12	Adequacy of communication equipment	Number of public telephone booths/ total population	Number/ 1,000,000 people
13	Adequacy of fire protection equipment	Number of fire apparatuses/ total population	Number/ 1,000,000 people
14	Adequacy of patient transportation equipment	Number of ambulances and heliambulances/ total population	Number/ 1,000,000 people
15	Capacity of local government finances	Financial ability **index** (standardized revenues/ standardized necessary expenditure of local government)	–
16	Capacity of local employment	Number of new job offers/ number of new job applications	%
17	Capacity of household budgets	Average saving rate (= savings/household income)	%
18	Capacity of labor force	Size of labor force/ total population	%

※ No. 01-08: prevention, No. 09-14: adaptation, No. 15-18: transformation

the names of the selected indicators and their associated assessment items. The urban indicators will be described in detail in Chap. 2.

1.5 Characteristics of This Book's "Resilience"

Let us introduce the summary of the characteristics of each guideline of resilience in terms of the above-mentioned concepts, frameworks, indicators, and so on. We reviewed some guidelines from UNISDR; United Nations University Institute of Advanced Studies (UNU-IAS); Resilient Alliance; World Bank; Arup, RPA and Siemens; Rockefeller Foundation and ARUP; Rockefeller Foundation; and Asian Cities Climate Change Resilience Network (ACCCRN) to make cities and communities resilient. The key points are summarized in Table 1.4.

Table 1.4 Characteristics of a variety of guidelines of resilience

	UNISDR	Bioversity International and UNU-IAS	Resilient Alliance	World Bank	Rockefeller Foundation and ARUP International Development	IIED and ACCCRN	This book
Name of guideline, year of publication	How to make cities more resilient A handbook for local governments leaders, 2012 (UNISDR 2012)	Indicators of Resilience in Socio-ecological Production Landscapes (SEPLs), 2013 (Bergamini et al. 2013)	Assessing Resilience in Social-ecological Systems Workbook for Practitioners Revised Version 2.0, 2010 (Resilient Alliance 2010)	Building Urban Resilience Principles, Tools, and Practice, 2013 (Jha et al. 2013)	City Resilience Framework, 2014 (The Rockefeller Foundation and ARUP International Development 2014)	Urban climate resilience: A review of the methodologies adopted under the ACCCRN initiative in Indian cities, 2013 (Sharma et al. 2013)	MRMR (Making cities and communities more resilient to multiple risks), 2018
Purpose	To make cities more resilient	Because of the dynamic nature and the complexity of interrelations between the elements of SEPLs, the indicators, jointly developed by Bioversity International and UNU-IAS, are designed to capture the different aspects that are entailed and essential for sustaining a resilient landscape (e.g., cultural, social, ecological, and agricultural)	Designed to assist in resolving specific resource issues and in developing and implementing management goals without compromising the resilience and integrity of the system as a whole	It summarizes the guiding principles, tools, and practices in key economic sectors that can facilitate incorporation of resilience concepts into the decisions about infrastructure investments and general urban management that are integral to reducing disaster and climate risks	This report presents the inclusive framework for articulating city resilience that the foundation was looking for, to underpin the City Resilience Index	This Working Paper aims to document and analyze the several methodologies adopted in the seven Indian ACCCRN cities: Surat, Indore, Gorakhpur, Shimla, Bhubaneswar, Mysore, Guwahati.	To make cities and communities more resilient to multiple risks for short and long term

(continued)

Table 1.4 (continued)

	UNISDR	Bioversity International and UNU-IAS	Resilient Alliance	World Bank	Rockefeller Foundation and ARUP International Development	IIED and ACCCRN	This book
				This handbook provides guidance on how to build urban resilience into critical infrastructure and the social realm by taking advantage of available methodologies, tools, and resources	This framework will form the basis of a tool that should enable all of us interested in city resilience to convene around a common understanding of that idea, and begin to "baseline" what matters most for making cities more resilient	The paper analyzes these methodologies and the overall process adopted in each of these cities for its potential for replication in *other cities in* India, and brings out the inherent challenges, gaps, and opportunities in achieving this	

						Drawing from these experiences, and with the aim of overcoming these challenges, this paper contributes recommendations on various stages of resilience planning exercises which would be beneficial to cities that plan to undertake such planning in the future	
Targets	Primarily for local government leaders and policy makers to support public policy, decision making, and organization as they implement disaster risk reduction and resilience activities	-	Practitioners	Urban planner	–	–	Mainly local government officials, but also the stakeholders and general citizens to develop policy collaboratively

(continued)

Table 1.4 (continued)

	UNISDR	Bioversity International and UNU-IAS	Resilient Alliance	World Bank	Rockefeller Foundation and ARUP International Development	IIED and ACCCRN	This book
Scale of the subject	Applying to cities and local governments, also applies to sub-national administrations of different sizes and levels, including at regional, provincial, metropolitan, city, municipal, township, and village level	While the origins of the term "SEPLs" are largely derived from an inclusive view of both mosaic landscapes and seascapes (Duraiappah et al. 2012), the term has recently been updated to explicitly encompass "socio-ecological production landscapes and seascapes," or SEPLS (Bergamini et al. 2013)	In contrast, an approach to managing natural resource systems that takes into account social and ecological influences at multiple scales, incorporates continuous change, and acknowledges a level of uncertainty has the potential to increase a system's resilience to disturbance and its capacity to adapt to change.	This handbook is a resource for enhancing disaster resilience in urban areas	City	City	City

			In contrast to attempting to control natural resources for stable or maximum production and short-term economic gain, a resilience approach assumes an uncertain and complex natural-resource context and aims to achieve sustainable long-term delivery of environmental benefits linked to human well-being				

(continued)

Table 1.4 (continued)

	UNISDR	Bioversity International and UNU-IAS	Resilient Alliance	World Bank	Rockefeller Foundation and ARUP International Development	IIED and ACCCRN	This book
Definition of resilience	Hazard × Vulnerability × Exposure/ Resilience or coping capacities = Disaster risk	The term SEPLs was coined to refer to mosaic production landscapes that have been shaped through long-term harmonious interactions between humans and nature in a manner that fosters well-being while maintaining biodiversity and ecosystem services (Gu and Subramanian 2012)	Resilience is fundamentally a system property	Resilience(UNISDR 2012): Resilience is the ability of a system, community, or society exposed to hazards to resist, absorb, accommodate to, and recover from the effects of a hazard in a timely and efficient manner	In the context of cities, resilience has helped to bridge the gap between disaster risk reduction and climate change adaptation	The definition of resilience as adopted under the ACCCRN program is "… the ability to absorb disturbances, to be changed and then to reorganise and still have the same identity (retain the same basic structure and ways of functioning)" and draws from literature on ecosystems and socio-ecological systems	We assume natural disasters and climate change, which influence each other, as one of the major human-caused external forces (risks or stresses), and these impacts various facets of socio-ecosystems

			It refers to the magnitude of change or disturbance that a system can experience without shifting into an alternate state that has different structural and functional properties and supplies different bundles of the ecosystem services that benefit people		Resilience focuses on enhancing the performance of a system in the face of multiple hazards, rather than preventing or mitigating the loss of assets due to specific events	The program thus views resilience as an ability of a system to not only withstand and resist climate circumstances, but also to recover and reorganize functions to prevent failures and irrevocable damages (Brown et al. 2012)	Along with the framework, we define the concept of resilient city and the scope of environmental resilience as following: resilient city is capable of adapting to multiple environmental risks, taking into consideration the relationship between the external forces (risks or stresses) and socio-ecosystems

(continued)

Table 1.4 (continued)

	UNISDR	Bioversity International and UNU-IAS	Resilient Alliance	World Bank	Rockefeller Foundation and ARUP International Development	IIED and ACCCRN	This book
						The concept of resilient systems as adopted under the ACCCRN framework is a system which avoids failures and has the ability to transform itself in the wake of changing climate. Rather than relying on the strength of individual components, resilient systems retain functionality through flexibility and diversifying functional dependence (Moench and Tyler 2012)	We assume that according to the degree of the external forces (risks or stresses), to make cities and communities more resilient, a combination of precautionary, adaptive, and transformation measures is required

Framework and principle	A disaster resilient city:	The indicators measure elements of SEPLs resilience that are, almost by definition, strongly interrelated	-	By breaking urban resilience down into four components, infrastructural, institutional, economic, and social, so that underlying issues can be addressed and capacity can be deepened	A resilient city is a city where there is or are...	Urban Climate Resilience Planning Framework (UCRPF) (Moench et al. 2011): The framework focuses on resilience more than adaptation and directs its attention to urban systems and their inherent interactions. The framework links the spatial, physical, and economic connections of direct climate impacts on these urban systems. The UCRPF focuses on the vulnerable population in urban locations and their marginalized subsistence that lacks secure access to services and depends on fragile urban systems.	A policy model is a hypothetical flow expressing the overall process of developing policy, based on the assumption that the state of implementation and preparation of resilience measures is governed by the three major elements of risks of external forces, vulnerabilities, and situations to be avoided

(continued)

Table 1.4 (continued)

	UNISDR	Bioversity International and UNU-IAS	Resilient Alliance	World Bank	Rockefeller Foundation and ARUP International Development	IIED and ACCCRN	This book
	Is one where disasters are minimized because the population lives in homes and neighborhoods with organized services and infrastructure that adhere to sensible building codes; without informal settlements built on flood plains or steep slopes because no other land is available	The practices and institutions that they describe can be grouped into four areas: (1) Ecosystems protection and the maintenance of biodiversity; (2) Agricultural biodiversity; (3) Knowledge, learning, and innovation; (4) Social equity and infrastructure. In the following section, each of these areas is discussed in greater detail		While this report addresses all components of disaster resilience, the focus is on the infrastructure and social aspects	1. Minimal human vulnerability: Indicated by the extent to which everyone's basic needs are met	UCRF (Urban Climate Resilience Framework): The key elements of the urban resilience framework are urban systems and social agents. Systems include ecosystems, infrastructure systems, institutions, and knowledge	It also provides the three types of indicator; urban indicators, citizen indicators, and administrative indicators which measures the state of each element of the policy model

	Has an inclusive, competent, and accountable local government that is concerned about sustainable urbanization and that commits the necessary resources to develop capacities to manage and organize itself before, during, and after a natural hazard event			◯Infrastructural resilience: Infrastructural resilience refers to a reduction in the vulnerability of built structures, such as buildings and transportation systems. It also refers to sheltering capacity, health care facilities, the vulnerability of buildings to hazards, critical infrastructure, and the availability of roads for evacuations and post-disaster supply lines. Infrastructural resilience also refers to a community's capacity for response and recovery.	2. Diverse livelihoods and employment: Facilitated by access to finance, ability to accrue savings, skills training, business support, and social welfare		Each element means that the indicators concerning risks from natural and social external force, indicators concerning vulnerabilities inherent to local communities and within the local government, indicators concerning anticipated situations that to be avoided, and indicators concerning the state of implementation and preparation of resilience measures capable of addressing these

(continued)

Table 1.4 (continued)

	UNISDR	Bioversity International and UNU-IAS	Resilient Alliance	World Bank	Rockefeller Foundation and ARUP International Development	IIED and ACCCRN	This book
	• Is one where the local authorities and the population understand their risks and develop a shared, local information base on disaster losses, hazards, and risks, including who is exposed and who is vulnerable			◯Institutional: Institutional resilience refers to the systems, governmental, and nongovernmental that administer a community	3. Adequate safeguards to human life and health: Relying on integrated health facilities and services, and responsive emergency services		Status report is used to diagnose the resilience of each city, by collecting together the results of measurement by indicators
	Is one where people are empowered to participate, decide, and plan their city together with local authorities and value local and indigenous knowledge, capacities, and resources			◯Economic resilience: Economic resilience refers to a community's economic diversity in such areas as employment, number of businesses, and their ability to function after a disaster	4. Collective identity and mutual support: Observed as active community engagement, strong social networks, and social integration		Policy scenario refers to the scenario developed in venues such as internal workshops within local government agencies and citizen conferences to implement policy, using the status report

	Has taken steps to anticipate and mitigate the impact of disasters, incorporating monitoring and early warning technologies to protect infrastructure, community assets, and individuals, including their homes and possessions, cultural heritage, environmental and economic capital, and is able to minimize physical and social losses arising from extreme weather events, earthquakes, or other natural or human-induced hazards			◯Social resilience: Social resilience refers to the demographic profile of a community by sex, age, ethnicity, disability, socioeconomic status, and other groupings, and the profile of its social capital. Although difficult to quantify, social capital refers to a sense of community, the ability of groups of citizens to adapt, and a sense of attachment to a place (Cutter et al. 2010)	5. Social stability and security: Including law enforcement, crime prevention, justice, and emergency management		

(continued)

Table 1.4 (continued)

	UNISDR	Bioversity International and UNU-IAS	Resilient Alliance	World Bank	Rockefeller Foundation and ARUP International Development	IIED and ACCCRN	This book
	Is able to respond, implement immediate recovery strategies, and quickly restore basic services to resume social, institutional, and economic activity after such an event.				6. Availability of financial resources and contingency funds: Observed as sound financial management, diverse revenue streams, the ability to attract business investment, adequate investment, and emergency funds		
					7. Reduced physical exposure and vulnerability: Indicated by environmental stewardship; appropriate infrastructure; effective land use planning; and enforcement of planning regulations		

					8. Continuity of critical services: Indicated by diverse provision and active management; maintenance of ecosystems and infrastructure; and contingency planning		
					9. Reliable communications and mobility: Indicated by diverse and affordable multi-modal transport systems and information and communication technology (ICT) networks; and contingency planning		

(continued)

Table 1.4 (continued)

	UNISDR	Bioversity International and UNU-IAS	Resilient Alliance	World Bank	Rockefeller Foundation and ARUP International Development	IIED and ACCCRN	This book
					10. Effective leadership and management: Involving government, business, and civil society, and indicated by trusted individuals; multi-stakeholder consultation; and evidence-based decision making		
					11. Empowered stakeholders: Indicated by education for all, and access to up-to-date information and knowledge to enable people and organizations to take appropriate action		

					12. Integrated development planning: Indicated by the presence of a city vision; an integrated development strategy; and plans that are regularly reviewed and updated by cross-departmental working groups		

(continued)

Table 1.4 (continued)

	UNISDR	Bioversity International and UNU-IAS	Resilient Alliance	World Bank	Rockefeller Foundation and ARUP International Development	IIED and ACCCRN	This book
Indicators	Essential 1: Institutional and Administrative Framework	(1) Ecosystem protection and the maintenance of biodiversity	Starting by using strategic questions and activities to construct a conceptual model of a social-ecological system that represents a place of interest, along with its associated resources, stakeholders, institutions, and issues	Disaster Resilience Indicators (Source: Adapted from Cutter et al. 2010—variables used to construct disaster resilience index by sub-components); it is probable that not all indicators will be collected during a single urban infrastructure project, but it is important to encourage better data collection.	Structure of the City Resilience Index: Categories:4, Indicators: 12, Sub-indicators: 48-54, Variables: 130-150	The SLF was modified by TARU by using proxy indicators to define capacities and vulnerabilities: 1 human (education), 2 social (social network and access), 3 financial (income stability; size of incomes, ratio of stable incomes to total, dependency ratio), 4 physical (lack of physical infrastructure access (water supply, sewerage, roads), 5 natural (water scarcity; separate from infrastructure/ floods)	The background of providing three types of indicators is that integration of expert knowledge, local knowledge, and living wisdom collected in these ways can increase environmental resilience effectively.

	Essential 2: Financing and Resources	Heterogeneity and multi-functionality of the landscape	Building on the conceptual model, the assessment guides the identification of potential thresholds that represent a breakpoint between two alternative system states and helps reveal what is contributing to or eroding system resilience		Qualities of resilient systems: Reflective, robust, redundant, flexible, resourceful, inclusive, integrated		○Administrative indicators: we identified 41 indicators concerning risks from natural and social external forces anticipated in past measures, 28 indicators concerning vulnerabilities inherent to local communities and within the local government, 24 indicators concerning anticipated situations that to be avoided, and 44 indicators concerning the state of implementation and preparation of resilience measures capable of addressing these

(continued)

Table 1.4 (continued)

	UNISDR	Bioversity International and UNU-IAS	Resilient Alliance	World Bank	Rockefeller Foundation and ARUP International Development	IIED and ACCCRN	This book
	Essential 3: Multi-hazard Risk Assessment—Know your Risk	Areas protected for their ecological and cultural importance	A resilience assessment can thus provide insight into developing strategies for buffering or coping with both known and unexpected change	◯Social resilience: age, educational equity, transportation access, etc.	◯Health & well-being		○Citizen indicators: 44 indicators are arranged to 16 indicators by integrating similar ones each other to facilitate understanding of citizen.
	Essential 4: Infrastructure Protection, Upgrading, and Resilience	Ecological links between landscape components for sustainable production			Minimal human vulnerability		○Urban indicators: The proposed indicators were then validated and carefully selected for data availability, simplicity, comparability, representa tiveness, and balance. Over 30 indicators were initially proposed, and 18 were carefully selected according to the criteria

	Essential 5: Protect Vital Facilities: Education and Health	Rate of recovery from extreme environmental and climate change-related stresses and shocks		◯Economic resilience: Housing capital, employment, income and equality etc.	Livelihoods & employment		
	Essential 6: Building Regulations and Land Use Planning	(2) Agricultural biodiversity: Maintenance, documentation, and conservation of agricultural biodiversity in a community/diversity of local food system			Safeguards to human life & health		

Table 1.4 (continued)

	UNISDR	Bioversity International and UNU-IAS	Resilient Alliance	World Bank	Rockefeller Foundation and ARUP International Development	IIED and ACCCRN	This book
	Essential 7: Training, Education, and Public Awareness	(3) Knowledge, learning, and innovation: Innovation in agricultural biodiversity management for improved resilience and sustainability/access and exchange of agricultural biodiversity/transmission of traditional knowledge from elders, parents, and peers to the young people in a community/cultural traditions related to biodiversity/number of generations interacting with the landscape/practices of documentation and exchange of local knowledge/use of local terminology or indigenous language/women's knowledge about biodiversity and its use		◯Institutional resilience: Mitigation, flood coverage, municipal services, political fragmentation, etc.	◯Economy & society		

	Essential 8: Environmental Protection and Strengthening of Ecosystems	(4) Social equity and infrastructure: Local resource governance/ autonomy in relation to land and resource management/gender/ social infrastructure/ health care/health risk			Collective identity & mutual support		
	Essential 9: Effective Preparedness, Early Warning, and Response			◯Infrastructure resilience: Housing type, shelter capacity, medical capacity, etc.	Social stability & security		
	Essential 10: Recovery and Rebuilding Communities			◯Community capital: Place attachment, political engagement social capital-religion, social capital-civic involvement, etc.	Finance including contingency funds		
					◯Infrastructure & environment		
					Reduced physical exposure		

Table 1.4 (continued)

	UNISDR	Bioversity International and UNU-IAS	Resilient Alliance	World Bank	Rockefeller Foundation and ARUP International Development	IIED and ACCCRN	This book
					Continuity of critical services		
					Reliable mobility & communications		
					◯Leadership & strategy		
					Effective leadership & management		
					Empowered stakeholders		
					Integrated development planning		

Example of application		Cuchillas del Toa Biosphere Reserve(Cuba)	—	Indonesia (3rd national program for community empowerment in urban areas), Timor-Leste (road climate resilience project), Nepal (modernization of Rani Jamara Kulariya irrigation scheme phase 1), Kiribati (Kiribati adaptation phase III), Colombia (First programmatic fiscal sustainability and growth resilience development policy loan), Papua New Guinea (Building a more disaster and climate resilient transport sector)	Surat (India), Concepcion (chili), New Orleans (USA), Semarang (Indonesia), Cali (Columbia), Cape Town (South Africa)	Surat, Indore, Gorakhpur, Shimla, Bhubaneswar, Mysore, Guwahati (India)	Sendai, Nonthaburi

As mentioned before, UNISDR not only focuses on disaster resilience, but also provides an overview of key strategies and actions needed to achieve sustainable development. The target is primarily the local government leaders and policy makers, and the guideline looks like a kind of check sheet consisting of ten major elements to which these targets respond.

Indicators of Resilience in Socio-ecological Production Landscapes (SEPLs), which UNU-IAS and Bioversity International developed, focus on sustaining a resilient landscape (e.g., cultural, social, ecological, and agricultural). It was operated in the form of focus group interview with stakeholders.

Resilient Alliance, which we mentioned before, was designed to assist in resolving specific resource issues and in developing and implementing management goals without compromising the resilience and integrity of the system as a whole. This is not necessary for a specific target but for practitioners in general, who fill in the worksheets consisting of multiple facets of the systems.

The World Bank's guideline is a resource for enhancing disaster resilience in urban areas which the definition referred to as UNISDR and the main target is urban planners. Based on their experiences of projects around the world, the guideline provides four components, namely infrastructural, institutional, economic, and social resilience so that the underlying issues can be addressed and capacity can be improved. The guideline also provides disaster resilience indicators consisting of four resilience components (social, economic, institutional, and infrastructural) and community capital, and encourages urban planners to collect better data for these.

The Rockefeller Foundation and ARUP International Development proposed City Resilience Framework and Index, which will form the basis of a tool that should enable all those interested in city resilience to develop a common understanding of that idea and begin to "baseline" what matters most for making cities more resilient. It forms the basis for the Rockefeller Foundation's 100 Resilient Cities to guide the development of City Resilience Strategies across the world. It defines a resilient city as helping to bridge the gap between disaster risk reduction and climate change adaptation and provides 12 comprehensive frameworks from minimal human vulnerability to integrated development planning. Its structure of the City Resilience Index (CRI) consists of four categories (health & wellbeing, economy & society, infrastructure & environment, and leadership & strategy), 12 indicators under these categories, and furthermore 48–54 sub-indicators under the indicators. These CRIs have been employed in the 100 Resilient Cities, and the selected Resilient Cities try to use them as far as possible.

Drawing from the experiences in seven Indian cities, International Institute for Environment and Development (IIED) and ACCCRN developed the guidelines to contribute recommendations at various stages of resilience planning exercises which would be beneficial to cities that intend to undertake such planning in the future. A definition of resilience is that it is an ability of a system to not only withstand and resist climate circumstances, but also to recover and reorganize functions to prevent failures and irrevocable damages; the concept of resilient systems as adopted under the ACCCRN framework is a system that avoids failures and has the ability to transform itself in the wake of changing climate. It provides several frameworks such as Urban Climate Resilience Planning Framework (UCRPF), which focuses on resilience more than adaptation, and directs its attention to urban systems and their inherent interactions. The framework links the spatial, physical, and economic connections of direct climate impacts on these urban systems. The UCRPF focuses on the vulnerable population in urban locations and their marginalized subsistence that lacks secure access to services and depends on fragile urban systems. The other framework is UCRF (Urban Climate Resilience Framework) in which the key elements are urban systems and social agents. Systems include ecosystems, infrastructure systems, institutions, and knowledge. The indicators listed to define capacities and vulnerabilities are: (1) human (education), (2) social (social network and access), (3) financial (income stability; size of incomes, ratio of stable incomes to total, dependency ratio), (4) physical (lack of physical infrastructure access, water supply, sewerage, roads), (5) natural (water scarcity; separate from infrastructure/floods).

As observed above, these guidelines have a variety of concepts, frameworks, indicators, and so on. Among these, this book discusses the following characteristics:

(i) We assume natural disasters and climate change, which influence each other, as two of the major anthropogenic forces (risks or stresses), impacting various facets of socio-ecosystems. On this premise, resilient city is capable of adapting to multiple environmental risks, taking into consideration the relationship between the external forces and socio-ecosystems; Environmental resilience should cover a broad range of issues, including energy, ecosystems, wastes, green urbanism, and others. Moreover, we assume that a combination of precautionary, adaptive, and transformative measures is required, based on the severity of the external forces.

(ii) Our primary target is local government officials who develop policies for making cities and communities resilient. Stakeholders and general citizens who collaboratively develop these policies are also included.

(iii) The policy model we proposed is a hypothetical model that describes the overall process of policy development, based on the assumption that the preparatory resilience measures and the extent of their implementation is governed by three major elements: risks of external forces, vulnerabilities, and anticipated situations to be avoided.

(iv) The policy model includes three types of indicators: urban, civil, and administrative indicators, which evaluate the state of each element: risks from natural and social external forces, vulnerabilities inherent to local communities and within the local government, anticipated situations to be avoided, preparation and implementation of resilience measures capable of addressing these.

(v) The background of providing the three types of indicators is that the integration of expert knowledge, local knowledge, and living wisdom collected in these ways can increase environmental resilience effectively. For administrative indicators, we identified 41 indicators concerning risks from natural and social external forces anticipated in past measures, 28 indicators concerning vulnerabilities inherent to local communities and within the local government, 24 indicators concerning anticipated situations that to be avoided, and 44 indicators concerning the state of implementation and preparation of resilience measures capable of addressing these. For the civil indicators, 44 indicators are reduced to 16 indicators by integrating indicators similar to each other to facilitate understanding of the citizens. For urban indicators, the proposed indicators were then validated and carefully selected for data availability, simplicity, comparability, representativeness, and balance. Over 30 indicators were initially proposed, and 18 were carefully selected according to the criteria.

We then discuss the empirical approaches using the concept, framework, and indicators of resilient policy model in following chapters.

References

Adger W (2000) Social and ecological resilience: are they related? Prog Hum Geogr 24:347–364
Allen KM (2006) Community-based preparedness and climate adaptation: local capacity building in the Philippines. Disasters 30(1):81–101
Bergamini N, Blasiak R, Eyzaguirre P, Ichikawa K, Mijatovic D, Nakao F, Subramanian SM (2013) Indicators of resilience in Socio-ecological Productiotan Landscapes (SEPLs), UNU-IAS policy report. UNU-IAS. Retrieved August 5, 2015, from https://www.bioversityinternational.org/fileadmin/user_upload/online_library/publications/pdfs/Indicators_of_Resilience_in_Socio-ecological_Production_Landscapes_SEPLs_1676.pdf
Brown A, Dayal A, Rumbaitis Del Rio C (2012) From practice to theory: emerging lessons from Asia for building urban climate change resilience. Environ Urban 24:531
Cutter SL, Burton CG, Emrich CT (2010) Disaster resilience indicators for benchmarking baseline conditions. J Homeland Security Emer Manag 7(1):51. http://regionalresiliency.org/library/Diaster_Resilience_Indicators_Susan_Cutter_et_al_2010_1281451159.pdf
Duraiappah AK, Nakamura K, Takeuchi T, Watanabe M, Nishi M (eds) (2012) Satoyama-satoumi ecosystems and human Well-being: socio-ecological production landscapes of Japan. United Nations University Press, Tokyo
Fujii S, Kume K, Matsunaga A, Nakano T (2012) Perspective of research on economic resilience. RIETI policy discussion paper series, 12-P-008 (in Japanese)
Godschalk D (2003) Urban hazard mitigation: creating resilient cities. Nat Haz Rev 4:136–143
Gu H, Subramanian SM (2012) Socio-ecological production landscapes: relevance to the green economy agenda. United Nations University Institute of Advanced Studies Policy Report
Holling CS (1973) Resilience and stability of ecological system. Annu Rev Ecol Syst 4:1–23
ICLEI-Local Government for Sustainability (2012) Preparing for tomorrow strategy 2012–2018, from http://archive.iclei.org/documents/LACS/Publicacoes/BH_Strategic_Plan_final_pdf
Ikeda S, Morioka T (1993) A concept and its methodology of risk analysis. Jpn J Risk Res 5(1):1–7. (in Japanese)
Ishihara Y, Nakamaru S (2007) Resilience – concepts, history and perspectives of studies. Ann Hiroshima Bunkyo Women's University, 42:53–81 (in Japanese)
Jha AK, Miner TW, Stanton-Geddes Z (eds) (2013) Building urban resilience: principles, tools, and practice. International Bank for Reconstruction and Development/The World Bank, Washington, DC
Manyena SB (2006) The concept of resilience revisited. Disasters 30(4):433–450
Mens, M. J. P., Klijn, F. de Bruijn, & K. M. van Beek, E. (2011). The meaning of system robustness for flood risk management. Environ Sci Pol, 14, 1121–1131

Moench M, Tyler S, Lage J (eds) (2011) Catalysing urban climate resilience – applying resilience concepts to planning practice in the ACCCRN program. Institute for Social and Environmental Transition, International, Boulder, p 306
Moench M, Tyler S (2012) A framework for urban climate resilience. Clim Dev 4(4):311–326
Mori A (2010) Points to remember in risk management in ecology – perspective of variation and non- equilibrium. Jpn J Ecol 60:337–348. (in Japanese)
National Resilience Promotion Office, Cabinet Secretariat (2014) Documents and minutes of national resilience (disaster reduction and mitigation) Advisory meeting 1st–15th. Retrieved August 1, 2014, from http://www.cas.go.jp/jp/seisaku/resilience
Newman P, Beatley T, Boyer H (2009) Resilient cities responding to peak oil and climate change. Island Press, Washington, DC
Norris FH, Stevens SP, Pfefferbaum B, Wyche KF, Pfefferbaum RL (2008) Community resilience as a metaphor, theory, set of capacities, and strategy for disaster readiness. Am J Community Psychol 41:127–150
Otto-Zimmermann K (ed) (2011) Resilient cities: cities and adaptation to climate change: proceedings of the global forum 2010. Springer, Dordrecht
Otto-Zimmermann K (ed) (2012) Resilient cities 2: cities and adaptation to climate change: proceedings of the global forum 2011. Springer, Dordrecht
Resilience Alliance (2010) Assessing resilience in social-ecological systems workbook for practitioners revised version 2.0. Retrieved August 5, 2015 from https://www.resalliance.org/resilience-assessment
Resilient. City.org. (2013) Resilience. Retrieved August 29, 2013, from http://www.resilientcity.org/index.cfm?id=11449
Resilient Alliance (2002) Key concepts – Resilience. Retrieved August 29, 2013, from http://www.resalliance.org/index.php/resilience
Sharma D, Singh, R., & Singh, R. (2013). Urban climate resilience: a review of the methodologies adopted under the ACCCRN initiative in Indian cities. Asian Cities Climate Resilience Working Paper Series 5.. Retrieved October 14, 2015, from http://pubs.iied.org/pdfs/10650IIED.pdf
Shiozaki Y, Kato T (2012) A definition of resilience and vulnerability in natural disaster and relevant areas. SEISAN-KENKYU 64(2):217–220
The Rockefeller Foundation and ARUP International Development (2014) City resilience framework.. Ove Arup & Partners International Limited. Retrieved August 5, 2015, from https://assets.rockefellerfoundation.org/app/uploads/20140410162455/City-Resilience-Framework-2015.pdf
Tobin GA (1999) Sustainability and community resilience: The holy grail of hazards planning? Environ Haz 1:13–25
Uchiyama I (2006) Health hazard, Health risk, environmental risk. In: Society for Risk Analysis Japan (ed) Risk dictionary (enlarged and revised edition). Hankyu Communications, Osaka, pp 42–46. (in Japanese)
UNISDR. (2012). How to make cities more resilient. A handbook for local governments leaders.. UNISDR. Retrieved August 29, 2013, from https://www.unisdr.org/files/26462_handbookfinalonlineversion.pdf
UNSIDR. (2005). Hyogo Framework for 2005–2015: Building resilience of nations and communities to disaster risk reduction.. Retrieved August 29, 2013, from http://www.unisdr.org/files/1037_hyogoframeworkforactionenglish.pdf
van Aalst MK, Cannon T, Burton I (2008) Community level adaptation to climate change: The potential role of participatory community risk assessment. Glob Environ Chang 18:165–179

Chapter 2
Assessment of City Resilience Using Urban Indicators in Japanese Cities

Shun Kawakubo, Kenshi Baba, Mitsuru Tanaka, Shuzo Murakami, and Toshiharu Ikaga

Abstract Conducting assessments using urban indicators based on public statistical information helps us understand the actual conditions and resilience of our cities and communities. Thereby it enhances our ability to resist, adapt to, and recover from devastating disasters. The outcomes of such assessments facilitate policymakers, government officers, businesses, citizens, and other stakeholders to detect weak points of the target city in comparison with other cities. In this light, various sets of indicators have been developed to assess the resilience of cities. Discussions on developing an international standard for resilience indicators have also started in a working group under the technical committee of the International Organization for Standardization. This chapter first briefly introduces the background of some important campaigns and movements for making cities more resilient. Next, two case studies of the assessment of city resilience in Japan are introduced. One is the case of introducing time-series assessments of a disaster-affected city to monitor the recovery process after a catastrophic earthquake using the CASBEE-City city-scale assessment tool. The other is the case of conducting the resilience assessment targeting major cities in Japan. These two studies show the importance of conducting the resilience assessment using urban indicators together with public statistical information.

Keywords Urban indicators · Resilient cities · Public statistical information · Recovery process · City-scale assessment · CASBEE

S. Kawakubo (✉) · M. Tanaka
Hosei University, Machida, Tokyo, Japan
e-mail: kawakubo@hosei.ac.jp

K. Baba
Tokyo City University, Yokohama, Kanagawa, Japan

S. Murakami
Institute for Building Energy Environment and Energy Conservation, Tokyo, Japan

T. Ikaga
Keio University, Yokohama, Kanagawa, Japan

M. Tanaka, K. Baba (eds.), *Resilient Policies in Asian Cities*,
https://doi.org/10.1007/978-981-13-8600-8_2

2.1 Importance of Assessing City Resilience Assessed by Public Statistical Information

Assessing the resilience of cities using appropriate indicators is essential when trying to make our cities more resilient, which enhances the ability to resist, adapt to, and recover from devastating disasters. According to Meadows (1998, p.1), "Indicators are a necessary part of the stream of information we use to understand the world, make decisions, and plan our actions." Information from appropriate indicators will support policymakers, government officers, businesses, citizens, and other stakeholders in understanding the actual situation and problems to be solved. In this light, many organizations are taking action to understand the resilience of cities. At the United Nations Office for Disaster Risk Reduction (UNISDR), the secretariat of the International Strategy for Disaster Reduction is promoting the Making Cities Resilient Campaign, supporting approximately 2500 cities globally (as of the end of 2014). The objective of this campaign is achieving resilient cities and communities through sufficient actions led by local governments to reduce disaster risks (UNISDR 2012). IBM and AECOM (2014) have developed the "Disaster Resilience Scorecard for Cities" for the UNISDR Making Cities Resilient Campaign, which enables cities to rate their resilience to disasters on a scale from 0 to 5. ICLEI-Local Governments for Sustainability (2010), one of the world's largest networks of cities, towns, and metropolises, is also promoting its Resilient Cities Program and discussing how to support local governments in tackling the issue of urban resilience. ISO/TC 268, a technical committee aiming to develop international standards for encouraging the development and implementation of holistic and integrated approaches to sustainable development in cities and communities, has also established a working group to consider how to assess cities and communities from perspectives such as city services, performance, sustainability, smartness, and resilience (International Organization for Standardization 2012). Indicators for assessing city resilience will be discussed in the working group, and a set of indicators will be published as an international standard in the future. When city resilience can be assessed using a set of indicators based on public statistical information, policymakers, government officers, businesses, citizens, and other stakeholders can more easily grasp the actual conditions of their cities and thereby detect local problems and take appropriate countermeasures.

Japan experiences disasters such as earthquakes, tsunamis, volcanic eruptions, typhoons, landslides, and floods. Many countermeasures based on past experience are thus taken to minimize damage and recover from such disasters. There are already some examples of utilizing indicators and public statistical information for assessing city resilience in Japan. This chapter introduces two recent examples of monitoring the actual conditions of target cities and assessing their resilience based on a set of indicators and public statistical information.

2.2 Case Study 1: Assessment of the Recovery Process After a Devastating Earthquake in Kobe, Japan

Earthquakes are one of the most terrifying natural disasters that cause catastrophic damage. Countries located near or above crustal plate boundaries frequently suffer damage due to large earthquakes. Japan is one such country, being located where the Pacific, Philippine Sea, Eurasian, and North American plates overlap. These plates converge, diverge, and grind against each other, and earthquakes result from the release of stress within the crust. Movement of these large plates has caused many devastating earthquakes resulting in widespread damage, both physically and mentally. However, disaster-affected cities have quickly recovered from earthquakes by implementing appropriate countermeasures leading to reconstruction of buildings, infrastructure, and social capital. From the perspective of life-cycle thinking, nothing remains the same forever; everything changes form, sometimes disappearing and sometimes reappearing. The same holds for the prosperity and decline of cities. Genuinely eternal sustainable development of cities and communities is thus impossible, but cities and communities can nonetheless become relatively more sustainable by overcoming disasters and adapting to changing environmental, social, and economic situations. When cities and communities wish to develop in a more sustainable way, reflecting upon the recovery process after disasters can be informative. Determining what kinds of countermeasures effectively contributed to disaster recovery is good preparation for future disasters. It is also beneficial to keep records of recovery efforts and processes and to share best practices with the world to contribute to achieving more resilient cities and communities around the world. This section introduces a methodology used for monitoring the recovery process after the occurrence of a devastating earthquake in Kobe, Japan, using a tool specifically developed for city-scale assessment.

2.2.1 *Kobe and the Great Hanshin–Awaji Earthquake*

Kobe is the largest city in Hyogo Prefecture. It is located in central Japan and has a population of over 1.5 million. It has a Shinkansen high-speed railway station in the city center and a major port in the south. Kobe has long been an economic and cultural center. Its population continuously increased from 1956, the year it became a government-designated city, until the time of the Great Hanshin–Awaji Earthquake, 17 January 1995 at 05:46:53 JST (16 January at 20:46:53 UTC). This devastating earthquake killed more than 6400 people, 4500 of whom were Kobe citizens. Figure 2.1 shows the location and an overview of the city.

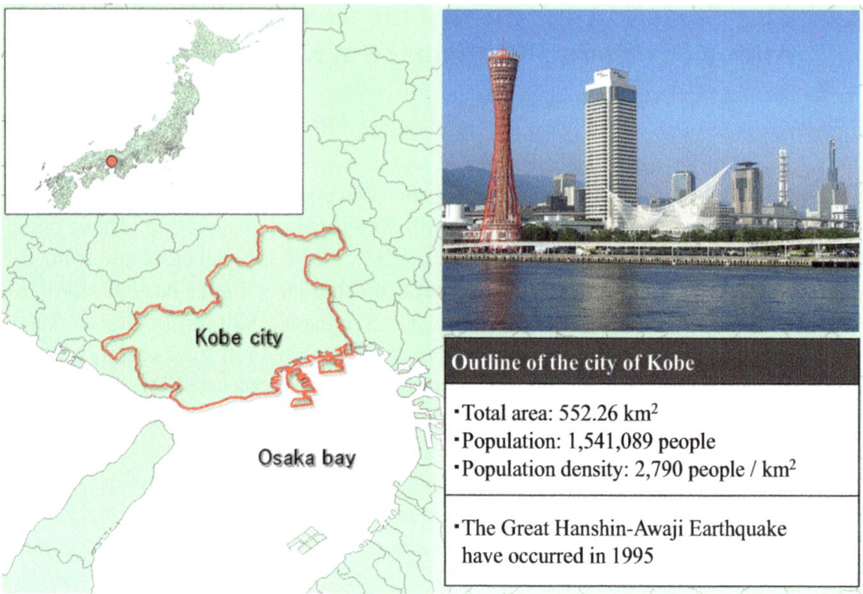

Fig. 2.1 Location of Kobe City, Hyogo Prefecture, and basic information

2.2.2 The CASBEE-City Comprehensive City Assessment Tool

The Comprehensive Assessment System for Built Environment Efficiency (CASBEE; Murakami et al. 2014) is a set of tools for comprehensively assessing individual buildings, multiple buildings including their surroundings district or urban blocks, and cities and communities as a whole. CASBEE for Cities (CASBEE-City; Murakami et al. 2011) is a tool developed specifically for city-scale assessment, and was used to monitor the recovery process in Kobe. This tool was developed by members of the Committee for the Development of Environmental Performance Assessment Tools for Cities with the support of the Office for Promotion of Regional Revitalization of the Cabinet Secretariat of Japan, the Ministry of Land, Infrastructure, Transport and Tourism, and local governments across the country. The tool and its manual are published by the Japan Sustainable Building Consortium (2012).

Figure 2.2 shows the assessment framework of the tool, which assesses target cities from two aspects: (1) quality (Q), including citizen quality of life and the environmental, social, and economic circumstances of the city, which quantify activities within it, and (2) environmental load (L), namely, greenhouse gas emissions resulting from human activities. Cities that achieve high quality of life while attaining lower greenhouse gas emissions are assessed as more sustainable within the CASBEE-City framework. Figure 2.3 and Table 2.1 show the assessment items and corresponding indicators, respectively. Figure 2.4 visualizes the assessment results.

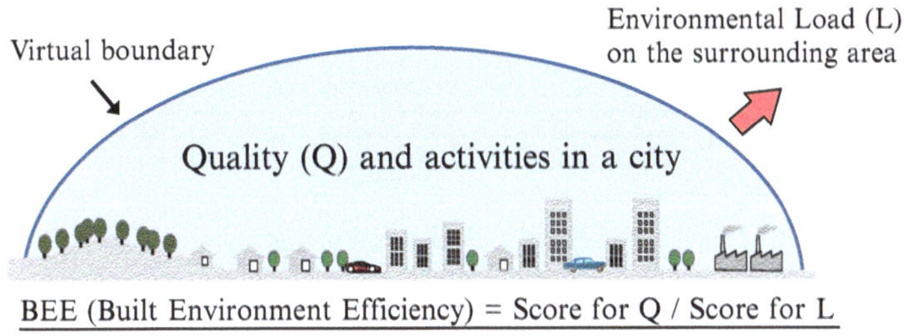

Fig. 2.2 Image of target city to be assessed by CASBEE-City tool

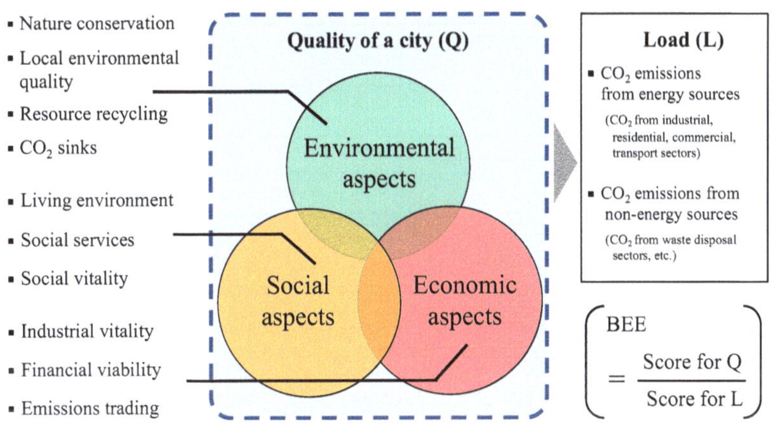

Fig. 2.3 Assessment items for the CASBEE-City tool

2.2.3 Assessment Results of Recovery Process Utilizing CASBEE-City

Comprehensive assessment of Kobe was conducted in terms of its past (pre-disaster), present (post-disaster), and future. Data from past to present were collected, and concrete future targets were also collected based on interviews with local government officers and other stakeholders. Finally, a time-series assessment of the city was conducted by inputting past-, present-, and future-estimated data into CASBEE-City. Figure 2.5 shows the time-series assessment for Kobe.

The graph shown in Fig. 2.5 is called a BEE chart. It shows Q scores on the vertical axis and L scores on the horizontal axis. Cities with high Q scores (achieving good environmental, social, and economic conditions) and low L scores (low greenhouse gas emissions per capita) are plotted in the upper-left side of the BEE chart, implying that they are more sustainable than other cities in the country. Tool users can compare multiple cities by simultaneously plotting their results. They can also compare estimated conditions of a target city at various times in a similar manner, as in this study, to visualize the recovery process of Kobe after its catastrophic

Table 2.1 Assessment items and corresponding indicators for the CASBEE-City tool

	Major and minor categories	Subcategory	Assessment indicator	Unit
Quality (Q)	Q1. Environmental aspects			
	Q1.1 Nature conservation	Q1.1.1 Ratio of green and water spaces	(Forest areas + major lake areas)/total land area	%
	Q1.2 Local environment	Q1.2.1 Atmosphere quality	Number of days in which the hourly concentration of photochemicals during the day is 0.12 ppm or higher (average)	Day
		Q1.2.2 Water quality	75% of average daily biochemical oxygen demand (BOD) in rivers	mg/l
	Q1.3 Resource recycling	Q1.3.1 Recycling rate of general waste	Recycling rate of general waste	%
	Q1.4 CO_2 sinks	Q1.4.1 CO_2 absorption by forests	(Current forest area unit of absorption)/adjusted population	t-CO_2/person
	Q2. Social aspects			
	Q2.1 Living environment	Q2.1.1 Quality of housing	Total floor area per dwelling unit	m^2
		Q2.1.2 Traffic safety	Number of traffic accidents/adjusted population	Number/1,000 people
		Q2.1.3 Crime prevention	Number of crimes recorded/adjusted population	Number/1,000 people
		Q2.1.4 Disaster preparedness	Number of disaster response hospitals per 100,000 persons	Number/100,000 people
	Q2.2 Social services	Q2.2.1 Adequacy of education services	(Number of students/number of teachers) at elementary and junior high schools	–
		Q2.2.2 Adequacy of cultural services	(Number of community centers + number of libraries)/land area of municipality	Number/10 km^2
		Q2.2.3 Adequacy of medical services	Number of doctors/adjusted population	Number/1,000 people
		Q2.2.4 Adequacy of childcare services	Number of childcare facilities/pre-school population (aged 0–4 years)	Number/100 people
		Q2.2.5 Adequacy of services for the elderly	Number of senior care facilities/senior population (aged ≥65 years)	Number/1,000 people
	Q2.3 Social vitality	Q2.3.1 Rate of population change due to births & deaths	(Number of births – number of deaths)/total population	%
		Q2.3.2 Rate of population change due to migration	(Number of move-ins – number of move-outs)/total population	%
	Q3. Economic aspects			
	Q3.1 Industrial vitality	Q3.1.1 Amount equivalent to gross regional product	(Agricultural output + value of manufactured goods shipments + sales of commercial goods)/adjusted population	1,000,000 yen/person
	Q3.2 Financial viability	Q3.2.1 Tax revenue	Tax revenues of the local government/adjusted population	10,000 yen/person
		Q3.2.2 Outstanding local bonds	Real debt service ratio	%
	Q3.3 Emissions trading	Q3.3.1 Emissions trading	Presence or absence of an emissions trading scheme	–
Environmental Load (L)	L1. CO_2 emissions from energy sources			
	L1.1 Industrial sector	CO_2 emissions from industrial sector/adjusted population		t-CO_2/person
	L1.2 Residential sector	CO_2 emissions from residential sector/adjusted population		t-CO_2/person
	L1.3 Commercial sector	CO_2 emissions from commercial sector/adjusted population		t-CO_2/person
	L1.4 Transportation sector	CO_2 emissions from transportation sector/adjusted population		t-CO_2/person
	L2. CO_2 emissions from non-energy sources			
	L2.1 Waste disposal sector and others	CO_2 emissions from waste disposal sector and others/adjusted population		t-CO_2/person

Note 1: Adjusted population = (daytime population + nighttime population)/2
Note 2: These assessment items and indicators are for CASBEE-City (2012 edition)

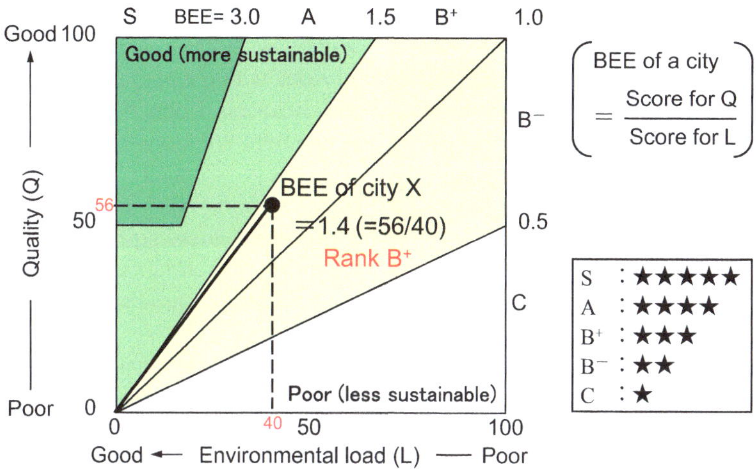

Fig. 2.4 Visualized assessment result (BEE chart)

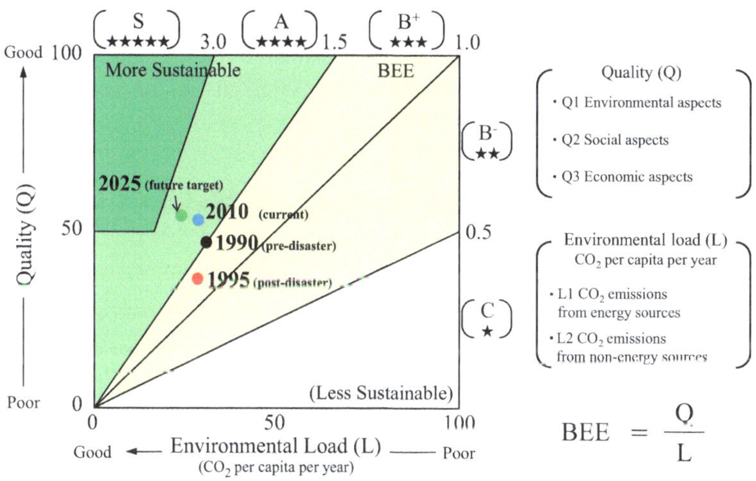

Fig. 2.5 Result of time-series assessment of the city of Kobe

earthquake. Regarding the city's time-series assessment results, quality (Q) of the city deteriorated after the 1995 Great Hanshin–Awaji Earthquake as compared with results from 1990. The environmental load (L) after the disaster slightly improved due to stagnation of economic activity in the disaster-affected areas. The decrease in Q exceeded the decrease in L, leading to a drop in BEE after the disaster. However, Kobe set reconstruction guidelines, and various stakeholders made enormous efforts to overcome the disaster. As a result, by 2010, conditions in Kobe had improved to levels exceeding those before the disaster. In 2015, the Japanese government selected Kobe as a model eco-city, and planning for sustainable development is ongoing. If current planning and measures are appropriately implemented, conditions of the city are expected to improve further by 2025. These outcomes were obtained through a city-wide dialogue held in Kobe with member companies of the Urban Infrastructure Initiative by the World Business Council for Sustainable Development, which has published several concrete solutions for achieving its ambitious sustainability targets (World Business Council for Sustainable Development 2013). Limited human and financial resources restrict what local governments can accomplish by themselves, but they can achieve ambitious targets by finding partners and stakeholders who play major roles in sustainable city development. This case demonstrates best practices for meta-governance in pursuing a more resilient and sustainable city.

It is necessary to review previously implemented measures that have led to successful recovery from past disasters, such as in the case of Kobe, and to share findings and best practices with other cities and communities to achieve a sustainable and resilient world. A comprehensive city assessment tool can support monitoring of the recovery process and identification of factors for successful recovery by comparing past and present conditions. The tool introduced in this case, CASBEE-City, is composed of various indicators for assessing the target city or community from multiple perspectives, but the set of indicators was not selected in terms of city resilience. The next section demonstrates assessment of major cities in Japan using a set of indicators specifically developed for assessing city resilience.

2.3 Case Study 2: Assessing Resilience of Major Cities in Japan

This section introduces a methodology for developing a set of indicators for assessing resilience of cities, and presents assessment results for major cities in Japan obtained though assessment using the developed set of indicators. There are 1750 municipalities in Japan, but this study focuses on the Tokyo metropolitan area and 18 government-designated cities with populations exceeding 500,000, because large cities are more vulnerable to disasters resulting in immense harm. Figure 2.6 shows the locations of the selected cities, and Table 2.2 lists some of their demographic characteristics.

2.3.1 Urban Indicators for Assessing City Resilience

Urban indicators were developed specifically for assessing city resilience by the following steps: Many indicators for measuring city resilience were proposed at first. The proposed indicators were then validated and carefully selected for data availability, simplicity, comparability, representativeness, and balance. Data

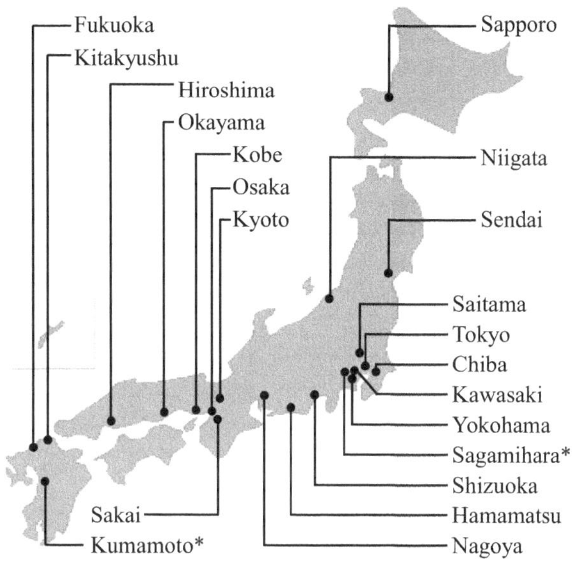

Fig. 2.6 Locations of the Tokyo metropolitan area and government-designated cities
Note: Sagamihara and Kumamoto are not assessed in this study due to lack of data

Table 2.2 Demographic information of cities assessed in the study

City	Prefecture	Total population	DID population	Households	Total area (km^2)
Sapporo	Hokkaido	1,913,545	1,846,399	885,848	1121.12
Sendai	Miyagi	1,045,986	931,677	465,260	788.09
Saitama	Saitama	1,222,434	1,126,138	503,126	217.49
Chiba	Chiba	961,749	869,241	406,309	272.08
Tokyo	Tokyo	8,945,695	8,945,695	4,540,746	621.98
Yokohama	Kanagawa	3,688,773	3,589,469	1,583,889	437.38
Kawasaki	Kanagawa	1,425,512	1,417,671	662,694	144.35
(Sagamihara)	Kanagawa	717,544	658,895	302,815	328.84
Niigata	Niigata	811,901	583,329	312,533	726.10
Shizuoka	Shizuoka	716,197	625,147	279,019	1411.82
Hamamatsu	Shizuoka	800,866	477,648	300,444	1558.04
Nagoya	Aichi	2,263,894	2,216,845	1,021,227	326.43
Kyoto	Kyoto	1,474,015	1,403,631	681,581	827.90
Osaka	Osaka	2,665,314	2,664,819	1,317,990	222.47
Sakai	Osaka	841,966	803,490	344,465	149.99
Kobe	Hyogo	1,544,200	1,440,411	684,183	552.83
Okayama	Okayama	709,584	478,993	296,790	789.91
Hiroshima	Hiroshima	1,173,843	1,012,198	512,907	905.41
Kitakyushu	Fukuoka	976,846	877,833	420,702	487.89
Fukuoka	Fukuoka	1,463,743	1,405,700	707,358	341.32
(Kumamoto)	Kumamoto	734,474	579,318	302,413	390.00

Note: DID, densely inhabited district
Note: Sagamihara and Kumamoto are not assessed in this study due to lack of data

availability is one of the most important criteria for selecting indicators, because even a theoretically interesting indicator is useless if its value cannot be calculated because of the lack of data. Simplicity is also important, as it helps indicator users to understand the actual condition of the target city. Comparability must be considered as well, because it is difficult to understand the strengths and weaknesses of a target city in comparison with other cities if the accuracy and definition of the data vary from area to area. Consideration of representativeness and balance is also essential when there are many candidate indicators for a single assessment factor. After developing the set of urban indicators in consideration of the above, the urban indicators were incorporated into three comprehensive city resilience indicators for assessing a city's (1) defensive capacity, which evaluates ability to prevent damage occurrence; (2) recovery capacity, which evaluates the ability to minimize damage after a disaster; and (3) learning capacity, which evaluates the ability to recover quickly from the disaster. Finally, the Tokyo metropolitan area were assessed and the other selected cities by using the developed indicators. The developed indicators are described in detail below.

Over 30 indicators were initially proposed, and 18 were carefully selected according to the aforementioned criteria. Table 2.3 lists the names of the selected indicators and their associated assessment items. Principal component analysis (PCA) was then applied to develop three comprehensive city resilience indicators and to calculate principal component scores. The 18 urban indicators were subjectively classified into three groups according to their defensive, recovery, learning capacities by considering features of the indicators. This is because when the 18 indicators were classified by PCA, the extracted components seemed to be correlated with area, population density, and geographical characteristics, not city resilience. Principal component scores were then calculated for each component by setting the number of factors to extract as 1. Table 2.4 shows the results of PCA as values for the principal component loadings.

2.3.2 *Resilience Assessment of Major Cities in Japan*

Standardized scores for each city for each urban indicator (average = 50, higher values indicate better conditions) were calculated and visualized on a line chart to detect strengths and weaknesses of the target city in comparison with the other assessed cities. As an example of the assessment results, Fig. 2.7 shows the case of the 18 urban indicators applied to Tokyo. Tokyo has clearly achieved good conditions for indicators 4 (prevention of accidents in an emergency situation), 10 (adequacy of medical professionals), 15 (capacity of local government finances), 16 (capacity of local employment), and 17 (capacity of labor force), while there is room for improvement for indicators 1 (prevention of fire breakout in dense areas) and 11 (adequacy of medical facilities).

Table 2.3 Assessment items and corresponding urban indicators

No.	Assessment item	Urban indicator	Unit
01	Prevention of fire breakout in density area	Number of fires/ population of densely-inhabited districts	Number / 1,000,000 people
02	Prevention of deterioration of industry production	Value of industry production/ /amount of water consumption	1,000,000 yen/ m^3/day
03	Prevention of evacuation route cutoff	Total length of roads with width less than 5.5 m/ total length of all type of roads	%
04	Prevention of accidents in an emergency situation	Number of traffic accidents/ total population	Number / 1,000,000 people
05	Prevention of delay in evacuation	Number of people requiring long-term care or support/ total population	%
06	Prevention of collapse of houses	Number of houses built before 1980/ total number of houses	%
07	Prevention of housing vacancies	Number of housing vacancies/ total number of houses	%
08	Prevention of uncomfortable room temperature	Number of houses with double-sash windows/ total number of houses	%
09	Adequacy of area for evacuation	Number of schools that can be used for evacuation/ total population	Number / 1,000,000 people
10	Adequacy of medical professionals	Number of doctors/ total population	Number / 1,000,000 people
11	Adequacy of medical facilities	Number of hospital beds/ total population	Number / 1,000,000 people
12	Adequacy of communication equipment	Number of public telephone booths/ total population	Number / 1,000,000 people
13	Adequacy of fire protection equipment	Number of fire apparatuses/ total population	Number / 1,000,000 people
14	Adequacy of patient transportation equipment	Number of ambulances and heliambulances/ total population	Number / 1,000,000 people
15	Capacity of local government finances	Financial ability index (standardized revenues/ standardized necessary expenditure of local government)	–
16	Capacity of local employment	Number of new job offers/ number of new job applications	%
17	Capacity of household budgets	Average saving rate (= savings/household income)	%
18	Capacity of labor force	Size of labor force/ /total population	%

Table 2.4 Result of principal component analysis (principal component loadings)

No.	Assessment item	Factor no. 1 (prevention)	Factor no. 2 (adaptation)	Factor No.3 (transformation)
01	Prevention of fire breakout in density area	0.489	–	–
02	Prevention of deterioration of industry production	0.368	–	–
03	Prevention of evacuation route cutoff	0.159	–	–
04	Prevention of accidents in an emergency situation	0.545	–	–
05	Prevention of delay in evacuation	0.813	–	–
06	Prevention of collapse of houses	0.851	–	–
07	Prevention of housing vacancies	0.646	–	–
08	Prevention of uncomfortable room temperature	0.446	–	–
09	Adequacy of area for evacuation	–	0.954	–
10	Adequacy of medical professionals	–	0.526	–
11	Adequacy of medical facilities	–	0.611	–
12	Adequacy of communication equipment	–	0.336	–
13	Adequacy of fire protection equipment	–	0.593	–
14	Adequacy of patient transportation equipment	–	0.655	–
15	Capacity of local government finances	–	–	0.315
16	Capacity of local employment	–	–	0.544
17	Capacity of household budgets	–	–	0.558
18	Capacity of labor force	–	–	0.823

Fig. 2.7 Example of assessment result using 18 urban indicators (case of Tokyo)

Information regarding the resilience of major cities in Japan can be obtained when principal component scores are plotted on a graph like that shown in Fig. 2.8. According to the obtained assessment results, cities near the Tokyo metropolitan area, such as Saitama, Kawasaki, and Yokohama, have achieved a high level of damage prevention. However, these cities should improve their recovery capacities to prevent catastrophic damage in a disaster. Kitakyushu shows the opposite trend compared with these three cities. Cities located in central Japan on the Pacific Ocean such as Hamamatsu, Nagoya, and Shizuoka (where the occurrence of a huge earthquake, the so-called Tokai earthquake, is predicted) have achieved high levels of learning capacity. The recovery capacities of Hamamatsu and Shizuoka are also high.

This section introduced a methodology for developing indicators for assessing city resilience and presented some results of applying them. The proposed assessments are only for the present year; however, it is important to continuously conduct assessments and compare them with past results, in order to understand whether measures and efforts taken in the city are successful.

2.4 Conclusion

This chapter discussed the importance of assessing city resilience based on an appropriate set of indicators and public statistical information through two case studies conducted in Japan. While these case studies focused on cities in Japan, the methodology for developing indicators and conducting assessments on city resilience is universally applicable and relevant in any region or country. Conducting assessments using appropriate indicators and monitoring trends in city resilience will contribute to achieving resilient and safe cities.

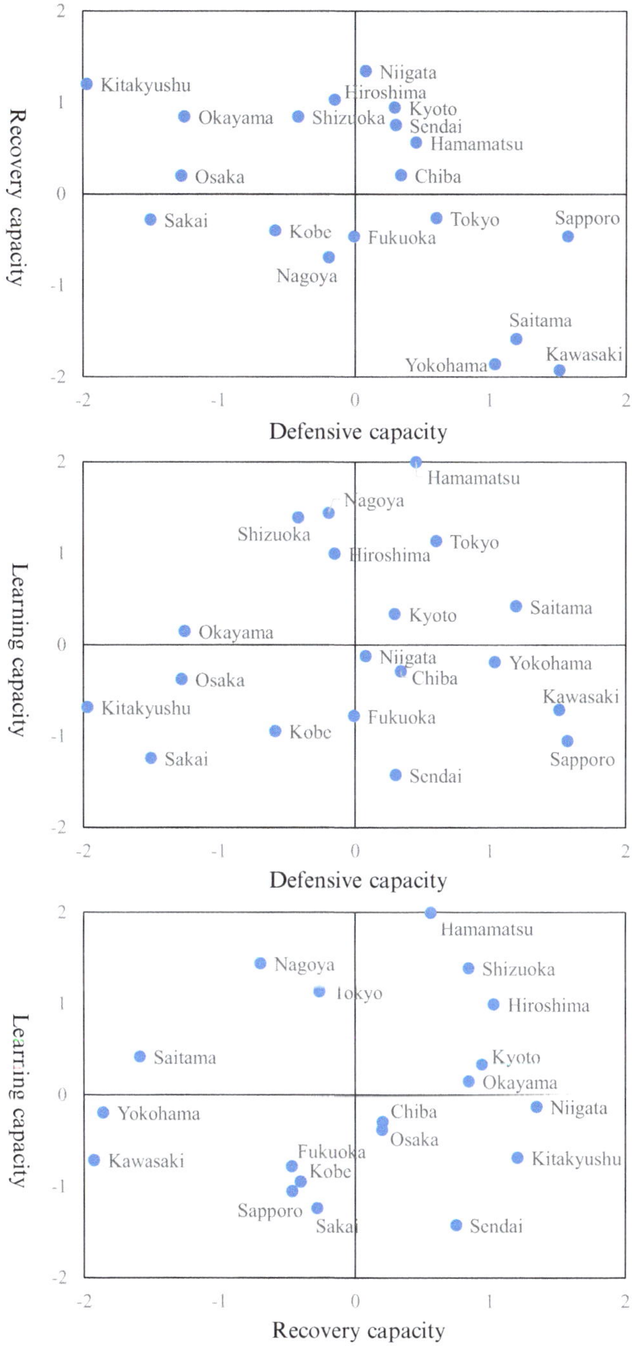

Fig. 2.8 Assessment results for city resilience

Acknowledgments The research results reported in this chapter were obtained in cooperation with members and officials of the Committee for the Development of Environmental Performance Assessment Tools for Cities; the Office for Promotion of Regional Revitalization of the Cabinet Secretariat of Japan; the Ministry of Land, Infrastructure, Transport and Tourism; and local governments across the country; and also, in cooperation with many other individuals involved in the project. Research outcomes of the first case study introduced in this chapter were obtained with the cooperation of the Kobe municipal government and member companies of the Urban Infrastructure Initiative by the World Business Council for Sustainable Development. We express our sincere gratitude to all persons concerned.

References

IBM and AECOM (2014) Working document – disaster resilience scorecard for cities,. [online]. http://www.unisdr.org/2014/campaign-cities/Resilience%20Scorecard%20V1.5.pdf. Accessed 24 Aug 2015

ICLEI-Local Governments for Sustainability (2010) Resilient City I ICLEI Global. http://www.iclei.org/index.php?id=36. Accessed 24 Aug 2015

International Organization for Standardization (2012) ISO – technical committees – ISO/TC 268 – Sustainable development in communities. http://www.iso.org/iso/iso_technical_committee?commid=656906. Accessed 24 Aug 2015

Japan Sustainable Building Consortium (2012) CASBEE for cities technical manual. Japan Sustainable Building Consortium, Tokyo

Meadows D (1998) Indicators and information systems for sustainable development: a report to the Balaton Group. Sustainability Institute, Hartland Four Corners

Murakami S, Kawakubo S, Asami Y, Ikaga T, Yamaguchi N, Kaburagi S (2011) Development of a comprehensive city assessment tool. Build Res Inform 39(3):195–210

Murakami S, Iwamura K, Raymond CJ (2014) CASBEE – A decade of development and application of an environmental assessment system for the built environment. Institute for Building Environment and Energy Conservation, Tokyo

United Nations Office for Disaster Risk Reduction (UNISDR) (2012) UNISDR-Welcome to making cities resilient. http://www.unisdr.org/campaign/resilientcities/. Accessed 24 Aug 2015

World Business Council for Sustainable Development (2013) A solutions landscape for Kobe Japan http://www.wbcsd.org/uiikobereport.aspx. Accessed 24 Aug 2015

Chapter 3
Civil Indicator: General Public's Cognitive Structure of Policies for Making Resilient Cities

Kenshi Baba, Kosuke Shirai, and Mitsuru Tanaka

Abstract This chapter attempts to clarify a cognitive structure of the general public in terms of policies for making cities resilient by analyzing data from web-based questionnaires in nine regions in Japan. The major findings are as follows: (i) Most respondents recognize the risk of earthquakes; most respondents regard "high percentage of elderly population, depopulation," as a vulnerability; most respondents regard "suspension of administrative activities" as a situation to be generally avoided. However, the scores of most indicators of risk and vulnerability differed among regions. (ii) The cognitive structure in which external forces risk, vulnerability, and situations to be avoided are assumed to determine the attitude to resilience measures was basically supported, but naturally differed among regions in certain details.

Keywords Questionnaire · Risk perception · Vulnerability assessment · Acceptability of measures

3.1 Introduction

This chapter introduces the results of an evaluation using civil indicators from data analysis derived from questionnaires responded to by the general public in nine Japanese regions. This data was subsequently used to evaluate citizens' external risk perception, vulnerability, and situations to be avoided. The impact of these parameters on evaluations of acceptability for each resilience measure was identified

K. Baba (✉)
Tokyo City University, Yokohama, Kanagawa, Japan
e-mail: kbaba@tcu.ac.jp

K. Shirai
Mitsubishi Research Institute, Inc., Tokyo, Japan
e-mail: kosuke_shirai@mri.co.jp

M. Tanaka
Hosei University, Machida, Tokyo, Japan
e-mail: mtanaka@hosei.ac.jp

M. Tanaka, K. Baba (eds.), *Resilient Policies in Asian Cities*,
https://doi.org/10.1007/978-981-13-8600-8_3

through verification of the validity of the structures hypothesized in Fig. 1.2 in Chap. 1. As mentioned in Chap. 1, the structure was generated from literature survey of diversified previous studies such as Adger (2000), Allen (2006), Bergamini et al. (2013), Cutter et al. (2008), Godschalk (2003), Hando and Kubota (2012), ICLEI-Local Government for Sustainability (2012), Jha et al. (2013), Lee et al. (2015), Longstaff et al. (2010), Norris et al. (2008), Sharma et al. (2013), Tobin (1999), UNISDR (2012), and van Aalst et al. (2008). We expect to form a clearer picture of the cognitive structure of citizens as recipients of risk-related information and highlight the differences between these constructs and how risks are framed by experts or those who disseminate the risk-related information. In the future, the authors hope to further the mutual understanding and social learning between actors through understanding discrepancies in framing between citizen cognitive structures (the recipients of risk information) and expert risk managers (the senders) by combining trial results for urban indicators mentioned in Chap. 2 and for civil indicators using separate datasets in Chap. 4.

3.2 Methodology

We designed questionnaire based on the vast relevant previous studies. As shown in Table 3.1, the survey was conducted online from February 7–12, 2015. A total of 3600 survey subjects took part in the survey according to the monitoring of the internet survey company. The participants lived in the Niigata and Kagoshima Prefectures, both of which are highly prone to sediment disasters [over the past 8 years, both prefectures have faced numerous mudflows, landslides, and cliff collapses (the Ministry of Land, Infrastructure, Transport and Tourism of Japan 2012)]; Kagawa and Aichi Prefectures, which are highly prone to droughts [over the last 20 years, both prefectures have been frequently affected by drought (the Ministry of Land, Infrastructure, Transport and Tourism of Japan 2012)]; Kyoto City and Kitakyushu City, which are highly prone to heat waves [the top two cities for patients falling victim to heat stroke in 2013 (National Institute of Environmental

Table 3.1 Survey overview

Survey period	February 7–11, 2015
Survey subjects	Local residents (men and women aged 20–60+) and government workers residing in the cities of Sapporo, Sendai, Kawasaki, Kyoto, and Kitakyushu or the prefectures of Niigata, Ehime, Kagawa, and Kagoshima
Method	Surveys returned via email or online
Questions	Perceived natural, social/external risks (41 indicators); community or household vulnerabilities (28 indicators); situations to be avoided (24 indicators); assessment of measures to respond to these (16 indicators); attributes of the individual, etc.
Responses collected	3600 (Approximately 400 responses × 9 regions)

Studies of Japan 2014)]; and Sapporo, which is located in a cold region with a climate significantly different to that of most other regions in Japan. Moreover, the authors chose Sendai City as a target area, as they are thought to exhibit high levels of interest and awareness in earthquakes and tsunamis, having been subject to the Great East Japan Earthquake.

Approximately, 400 subjects living in a total of nine regions including specific regions affected by frequent natural disasters (such as drought, landslides, cold waves, blizzards, and earthquakes) or health hazards (e.g., heat stroke) were surveyed and other regions were used for comparison. Responses were grouped equally by gender and age (i.e., five categories from those aged 20 to those 60+ in 10 year increments). Where there were ˂40 respondents in their 20s or 60+, we increased the number of respondents in the adjoining groups (for example, we increased the number of male respondents in their 30s when there was an insufficient number of males in their 20s) so that the number of respondents in each city reached a designated threshold.

Questions were chosen to identify perceived natural and social/external risks (41 indicators), vulnerabilities within local communities and households (28 indicators), and situations to be avoided (24 indicators). These were identical to those used in a local government survey that aimed to measure administrative indicators, as discussed in Chap. 5. Respondents were asked the acceptability of 16 measures to make a city more resilient as a preference of the following three types: the measures should be implemented via self-support/by individuals (I will do what I can by myself), by community support (I participate in local activities run by neighborhood committees and associations), or by public support (I support/accept government measures and activities).

The survey also contained numerous questions aimed at eliciting individual attributes such as the resident's own experience regarding personal damages/disasters (or lack thereof), social capital, number of years residing at their current residence, living arrangements, and future housing-related plans.

Respondents indicated they and their family's experience regarding particular personal damages by selecting from the following options: (1) experienced the shutting off or limiting of essential utilities such as water and electricity; (2) suffered damage to property such as vehicles or homes; (3) experienced living in an evacuation shelter; (4) suffered physical injury or sickness; (5) suffered health problems including the aggravation of a health condition; (6) suffered damages resulting in the disruption or cessation of work or business; or (7) no such experience.

Social capital was described according to two types, with the first addressing forms of social involvement or organizations with which respondents most frequently engage. Respondents were directed to select all applicable options from: (1) interactions with neighbors; (2) friends or acquaintances; (3) relatives; (4) coworkers; (5) local groups; (6) sports, hobbies, or recreational groups; (7) volunteer groups; (8) civic or consumer action groups; (9) government groups; (10) religious groups; (11) other; (12) none in particular; and (13) no response. The second type of social capital focused on the respondent's activities within the local community and the respondents were directed to select all applicable options from the following: (1) greet neighbors; (2) chat and consult with neighbors; (3) know where elderly people who live alone are located; (4) can resolve disputes within the neighborhood; (5) regularly participate in local crime prevention or clean-up activities; (6) know locations in the neighborhood that may be hazardous during a disaster (cliffs, low-lying areas, etc.); (7) have participated in training to set up evacuation shelters; (8) participate in festivals, bazaars, and events; and (9) none in particular.

The analysis method used to clarify people's acceptability to resilience measures, such as those hypothesized in Fig. 1.2 in Chap.1, each indicator, i.e., risk perception of natural and external social forces (41 indicators), assessment of vulnerabilities in local community and the home (28 indicators), and evaluations of situations to be avoided (24 indicators), was a simple tabulation of acceptability evaluations (the desirability of having self-support, community support, or public support as the implementing actor; 16 indicators) toward resilience measures that were applicable responses. After developing an outline of aggregates by region and obtaining the dispersion analysis results, the authors examined the cognitive structures using covariance structure analysis.

3.3 Results and Observations

3.3.1 Perceptions of Risk

Figure 3.1 shows the average evaluations aligned on a 5-point scale (ranging from "not perceiving that it could be a crisis" to "perceiving that it could be a fairly serious crisis") for 41 indicators of risk perception of natural forces and external social forces by region. Moreover, either ∗∗ or ∗ denoted indicators that were found to have significant variations of $p < 0.01$ or $p < 0.5$, respectively, after the relevant analysis. Broadly speaking, risk awareness for "earthquakes" was particularly high, followed by high awareness of the risks of "aging of population," "transportation accidents,"

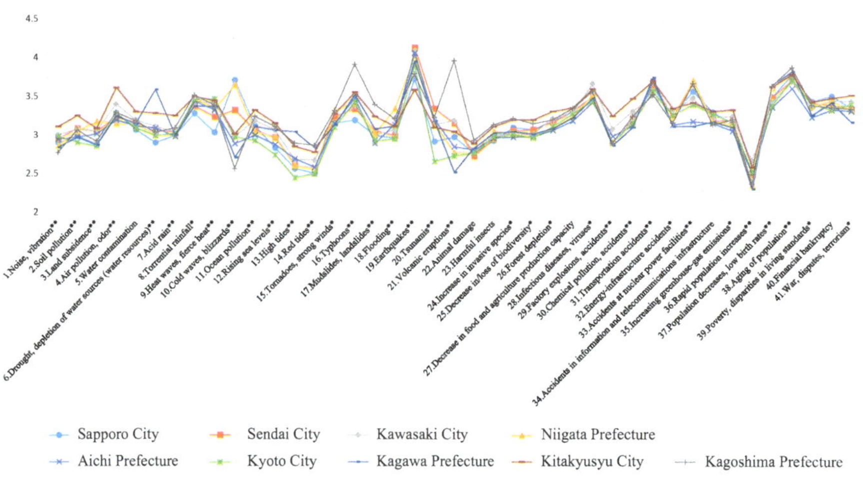

Fig. 3.1 Risk perception caused by external forces (risk/stress)

"population decreases, low birth rates," and "infectious diseases, viruses." When separating the average values for all indicators by region, it was clear that Kitakyushu City, Kagoshima Prefecture, and Kawasaki City showed the highest values in descending order, while Kyoto City and Aichi Prefecture were particularly low.

Only 6 out of the 41 indicators showed no significant variation: "water contamination," "animal damage," "harmful insects," "decrease in food and agriculture production capacity," "accidents in information and telecommunications infrastructure," and "financial bankruptcy"; but variations by region were observed. Separated by region, it is clear that cold regions, such as Sapporo City, Sendai City, and Niigata Prefecture, showed high risk awareness of "cold waves, blizzards," while the residents of Kyoto and Kitakyushu City exhibited high risk awareness of "heat waves, fierce heat." Moreover, while all regions showed high scores for "earthquake" risk, Sendai City had the highest risk awareness for "tsunamis," while regions that would not be exposed to a tsunami, such as Kyoto and Sapporo City, did not show high levels of awareness. Kitakyushu City showed high levels of awareness for pollution-related indicators, such as "noise, vibration," "soil pollution," and "air pollution, odor." Kawasaki City also exhibited high levels of awareness for "air pollution, odor" and the highest awareness of "infectious diseases and viruses," suggesting that regions with dense populations show high risk awareness. Moreover, high awareness was observed in Kagawa Prefecture for "drought, depletion of water sources (water resources)" and in Kagoshima Prefecture for "typhoons" and "volcanic eruptions," possibly reflecting the geographical contexts of these regions.

Moreover, t-testing and chi-square testing for relationships with individual attributes were performed.

With regards to gender, statistically significant differences at $p < .01$ were noted for 33 risks, with differences at $p < .01$ noted for 3 risks, while only 5 risks – "torrential rainfall," "typhoons," "energy-infrastructure accidents," "transportation accidents," and "population decreases, low birth rates" – showed no meaningful differences between men and women. Men had a higher perception of risk related to "population decreases, low birth rates," while all other risks were perceived more highly by women.

With respect to age, chi-square testing was conducted after dividing the sample into age cohorts between 20 and 70. Significant differences at $p<.01$ were noted for all risks except for those relating to flooding and tsunamis. Perceptions of all risks tended to increase with increasing age.

T-testing was performed after dividing the sample between people who had experienced damages and people who had not, producing significant differences at $p < 0.05$. Significant differences were observed against all experiences for the risks of "land subsidence," "torrential rainfall," "cold waves, blizzards," "flooding," "earthquakes," "tsunamis," and "accidents at nuclear power facilities." Significant differences were observed in relation to 23 risks for experiences regarding damage, excluding evacuation, such that a respondent's experience (or lack thereof) regarding damages/disasters significantly affected their perceptions of more than half of the relevant risk categories. No observable difference according to experience in relation to the risk of rapid population increases was obtained.

3.3.2 Assessment of Vulnerability

Figure 3.2 shows the average response values aligned on a 5-point scale (ranging from “not vulnerable at all” to “very vulnerable”) for 28 indicators used to assess vulnerability. Moreover, the analysis results are marked with either ∗∗ or ∗ for indicators with significant variations of $p < 0.01$ or $p < 0.5$, respectively. In particular, respondents indicated higher levels of vulnerability where there were “high percentage of elderly population, depopulation,” “high percentage of low-income earners, poor employment conditions,” or “presence of earthquake zones or fault lines.”

Only 2 out of the 28 indicators exhibited no significant variation by region: “high number of single-person households” and “presence of political conflict”; and similar to that of external risk awareness, variation by region was observed. While all regions showed high scores for “earthquakes,” Sendai City had the highest for “presence of earthquake zones or fault lines,” while Kitakyushu and Sapporo City showed low scores. Moreover, Kagoshima Prefecture showed an extremely high score for “presence of active volcanoes” and Kagawa Prefecture for “lack of usable water resources” compared to those of other regions, reflecting their geographical context, as was the case for awareness of certain external risks. In addition, Kagoshima Prefecture exhibited the highest scores for “high number of vacant homes,” “high percentage of low-income earners, poor employment,” and “high percentage of elderly population, depopulation,” suggesting a strong awareness among its inhabitants of regional depopulation issues. In contrast, Kawasaki City (which is close to urban centers) showed low scores for these vulnerability indicators. Another notable trend was that with many traditional machiya (town-houses) remaining in Kyoto City, it showed the highest score for “high number or density of wooden homes.”

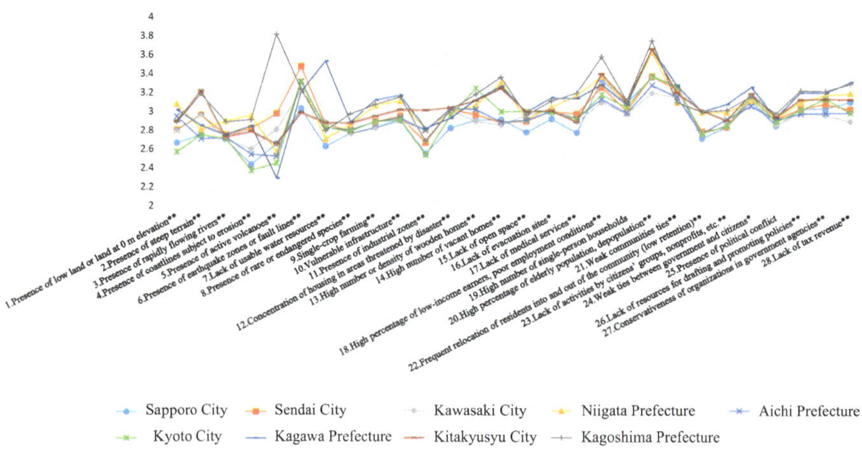

Fig. 3.2 Vulnerabilities deemed present within local communities and households

Results of the t-testing and chi-square testing with regard to relationships with individual attributes were as follows. Differences between men and women's perceptions were statistically significant at $p < .01$ for vulnerabilities associated with the "presence of rapidly flowing rivers," "presence of active volcanoes," "presence of earthquake zones or fault lines," "lack of usable water sources," "lack of evacuation sites," "lack of medical services," "higher number of single-person households," "weak ties within communities," and "lack of activities by citizens' groups, nonprofits, etc." Differences were significant at $p < .05$ in relation to only the factor of "high percentage of low-income earners, poor employment conditions." Women generally assessed these vulnerabilities higher scores than men.

Results of the chi-square testing with regards to age brackets produced significant differences at $p < .01$ among 17 perceived vulnerabilities and at $p < .05$ among 6 vulnerabilities. No significant differences were observed among the vulnerabilities associated with "lack of useable water sources," "presence of industrial zones," "lack of open space," "lack of evacuation sites," or "lack of medical services." Awareness of multiple vulnerabilities decreased with increasing age of the respondent.

The t-testing conducted after dividing the sample between people who had (and had not) experienced a relevant disaster/type of damage produced significant differences at $p < .05$ for a number of risks. Each experience regarding damage/disaster was associated with significant differences in perception of the "presence of earthquake zones or fault lines" and "high number or density of wooden homes." Experiencing six of the relevant types of damage resulted in significant differences in perceptions related to "high percentage of low-income earners, poor employment conditions," "high percentage of elderly population, depopulation," and "conservativeness of organizations in government agencies." In contrast, seven vulnerabilities were not perceived differently for all experiences regarding damages/disasters, a larger number than observed for external risks. Thus, experiencing damage/disaster had a relatively low impact on a respondent's vulnerability assessment.

The results of the chi-square testing with regard to respondent attributes related to housing arrangements were obtained. Vulnerabilities that showed significant differences at $p < .01$ were observed according to the number of years in a current residence including the "presence of earthquake zones or fault lines," "high number or density of wooden homes," "high percentage of elderly population, depopulation," and "lack of resources for drafting and promoting policies." Significant differences at $p < .05$ were also noted for vulnerabilities related to "single-crop farming," "vulnerable infrastructure," "high number of vacant homes," "lack of activities by citizens' groups, nonprofits, etc.," and "weak ties between government and citizens." Similar to the perception of external risks, fewer years spent in one's current residence was associated with a lower awareness of vulnerabilities and vice versa. Variance analysis on the respondent's intentions with regard to their future living arrangements produced significant differences at $p < .01$ for 12 vulnerabilities and at $p < .05$ for 8 vulnerabilities. These intentions affected a respondent's awareness of many vulnerabilities. Generally, an inverse relationship was observed whereby a stronger preference for remaining in one's current residence was associated with a lower vulnerability awareness. Differences in the respondent's current living arrangements produced significant differences at p<.01 for perceptions of vulnerabilities associated with “high number or density of wooden homes,” “high number of vacant homes,” “high percentage of low-income earners, poor employment conditions,” “high percentage of elderly population, depopulation,” and “weak ties within communities.” A difference at p<.05 was noted for vulnerability associated with “lack of tax revenue.” Respondents living in detached (rented) homes showed a higher awareness of vulnerabilities, while those living in apartment buildings were particularly cognizant of “weak ties within communities.”

3.3.3 Evaluation of Anticipated Situations to Be Avoided

Figure 3.3 shows the average response values for 24 indicators related to anticipated situations to be avoided scored on a 5-point scale (ranging from “not to be avoided at all” to “to be avoided absolutely”). In particular, respondents indicated higher levels of situations to be avoided for “cut-off of food or lifeline supply,” “cut-off of energy supply,” “long-term worsening of food conditions,” and “long-term instability in energy supply,” and these indicators were rated particularly highly.

Only 3 of the 24 indicators showed significant variation: “chronic worsening of public safety,” “degeneration of local culture/traditions,” and “long-term negative impact on ecosystems.” In most cases, no significant regional variation was observed, in contrast to external risk and vulnerability awareness. Separated by region, Kitakyushu City showed a high score while Kagoshima Prefecture exhibited a low score for “chronic worsening of public safety” and Kyoto City was high for

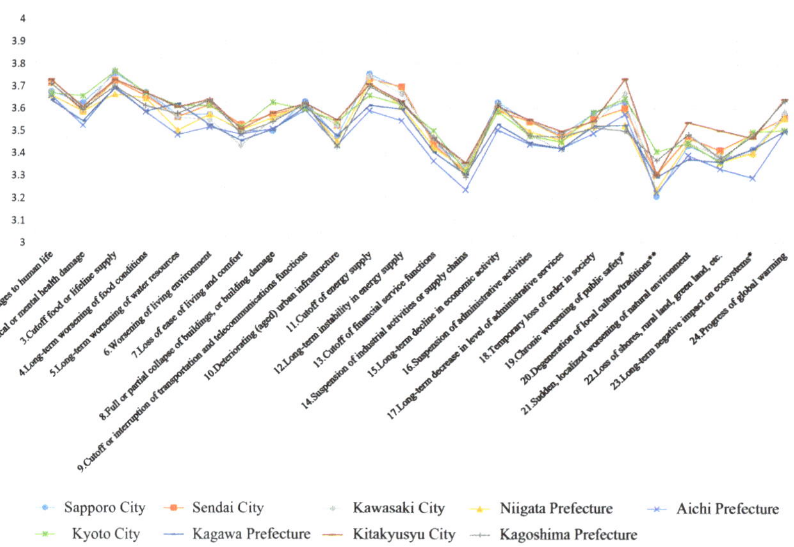

Fig. 3.3 Situations to be avoided that were deemed to require preventative measures

"degeneration of local culture/traditions." Aichi Prefecture showed a low score for "long-term negative impact on ecosystems" (not necessarily evaluated highly as a situation to be avoided), having hosted the tenth meeting of the Conference of the Parties for the Convention on Biological Diversity (COP10), where the Aichi Targets were adopted. Aichi Prefecture exhibited slightly low scores for all indicators.

The results of the t-tests and chi-square tests with regard to individual attributes were as follows. The t-testing for gender-based differences produced significant differences at $p<.01$ for 19 situations and at $p<.05$ for two situations. No significant differences were observed among situations involving "deteriorating (aged) urban infrastructure," "cut-off of energy supply," or "suspension of industrial activities or supply chains." For all situations showing significant differences, women tended to rate these situations to be avoided more highly than men, as was the case with perceptions of external risks and vulnerabilities.

The results of the chi-square testing with regard to age produced significant differences at $p<.01$ for situations involving "deteriorating (aged) urban infrastructure," "cut-off of energy supply," "long-term instability in energy supply," "temporary loss of order in society," "degeneration of local culture/traditions," "long-term negative impact on ecosystems," and "increase in global warming." Significant differences at $p<.05$ were observed for situations involving "long-term physical or mental health damage," "long-term decline in economic activity," and "loss of shores, rural land, green land, etc." Most situations were scored lower by respondents in their 20s.

The results of t-testing with regards to a respondent's experience of damages/disasters showed no significant differences among assessments of any situation associated with a respondent's experience of being evacuated from his/her home. The remaining six types of experiences produced significant differences for all situations to be avoided.

The sample population was subjected to t-testing after being divided into respondents with and without neighborly interactions and engagement. The associated results indicate significant differences among respondents who recorded the highest 10 items of social interaction/organizational membership in terms of their assessment of "long-term decrease in the level of administrative services," "degeneration of local culture/traditions," "long-term negative impact on ecosystems," and "increase in global warming." Conversely, social interaction and organizational membership did not produce significant differences with regard to assessments of situations involving the disruption of financial services. Nevertheless, having more than six types of social interaction/membership showed a discernible effect on how respondents scored all situations to be avoided. Regarding engagement in local community, being "able to resolve disputes in the neighborhood" did not produce significant differences in relation to assessments of any situations to be avoided, similar to that of perceptions of external risks. Global warming was the one situation for which a significant difference was observed for the eight community engagement items. Perceptions least affected by community engagement were those relating to the loss of livability and disruption to financial services. However, more than four types of neighborly interaction/engagement produced significant differences in how all situations to be avoided were assessed, highlighting the impact of social capital.

The results of chi-square testing with regard to housing-related attributes were assessed. Assessments of all situations to be avoided were unaffected by differences in the number of years residing at one's current residence and can be considered negligible as a variable. Variance analysis of the effects of future intentions with regard to current place of residence showed that these intentions have a large impact on assessments of situations to be avoided, as was the case with perceptions of external risks and vulnerabilities, producing significant differences at $p < 0.01$ for six situations and $p < 0.05$ for 14 situations.

3.3.4 Evaluation of Acceptability of Resilience Measures

Figure 3.4 shows the ratio of responses to questions regarding the 12 measures aimed at responding to risks (resilience measures), vulnerabilities, and situations to be avoided. Specifically, the survey asked whether respondents believed the measures should be performed via self-help/by individuals (I will do what I can by myself), in cooperation with others (I participate in local activities run by neighborhood committees and associations), or by the government (I support/accept government measures and activities). For many measures, more than half of the respondents indicated government intervention was the most appropriate approach. This was particularly the case for "reviewing and concentration of urban functions" and "the special deregulation zone system." Community support was rated as the second most preferred method after public support for almost all measures assessed, though

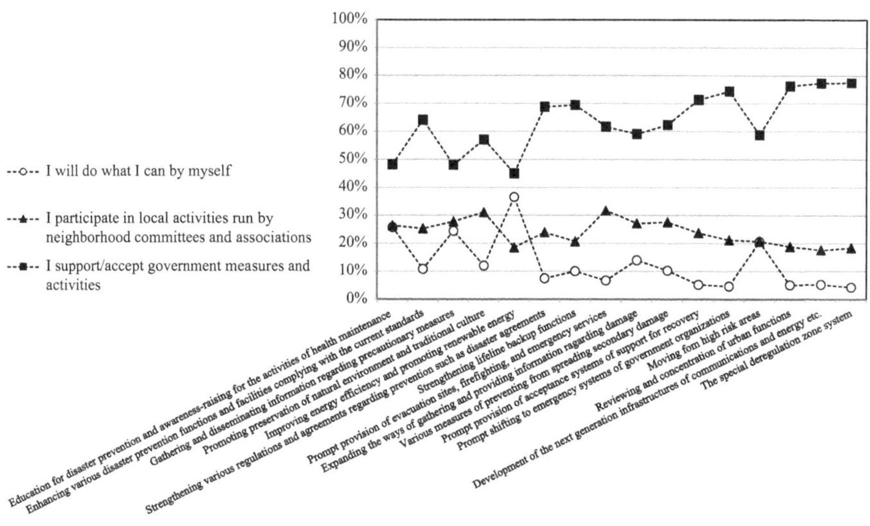

Fig. 3.4 Acceptability of the proposed resilience measures

these were surpassed by a preference for self-support for "improving energy efficiency and promoting renewable energy" and "moving from high risk areas."

Figure 3.5 shows the average values of the respondents selecting self, community, and public support by counting the numbers selected for each type individually and separated by region. Significant variation ($p = 0.020$) was only found between the regions for self-support, where this tendency was high in Sapporo City and low in Kagawa Prefecture.

3.3.5 *Examining the Cognitive Structure*

Below, we used covariance structure analysis to consider what impacts a citizen's perceptions of external risks, vulnerabilities, and situations to be avoided have on how they assess various methods for implementing resilience measures.

Firstly, the 5-point scale responses for indicators relating to external risks (41), vulnerabilities (28), and situations to be avoided (24) were subjected to a factor analysis using the maximum likelihood method and promax rotation. The five items with largest factor loading within the extracted factors (or the number of items for factors constructed using less than five items) were used as observed variables to construct the factors. Subsequently, a path was drawn according to the partial correlation coefficients of each factor. Although factor analysis was conducted for each region and the extracted factors used to produce the model, it proved difficult to compare regions using a uniform model due to differences in extracted factors and constitutive observable variables, rendering it invalid. Therefore, the authors selected factors extracted for most regions to create a secondary model using communal factors that were common across all regions.

The final factors used and the variables that comprise those factors are shown in Table 3.2. By testing the suitability of each factor, Cronbach reliability coefficients were obtained and are listed. In most cases, these reliability factors exceeded 0.8,

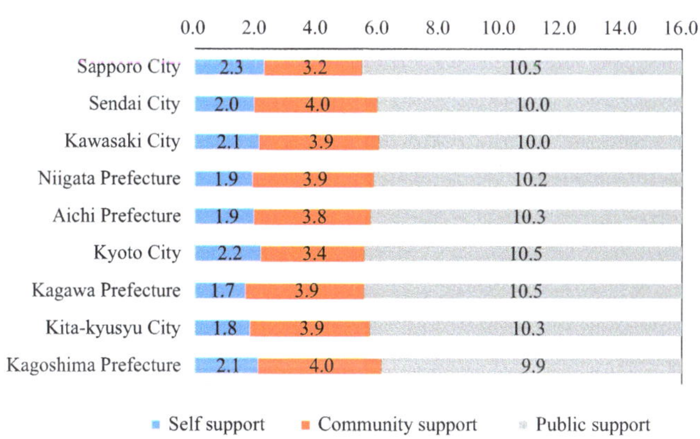

Fig. 3.5 Acceptability toward the resilience measures

Table 3.2 Factors and observed variables

	Factor	Cronbach's reliability coefficient	Relevant observed variables
External risks	Social risks	.839, .849	Population decreases, low birth rates
		.858, .843	Aging of population
		.914, .850	Financial bankruptcy
		.848, .812	Decrease in food and agriculture production capacity
		.855, .808	Poverty, disparities in living standards
	Disaster risks	.825, .828	Torrential rainfall
		.827, .826	Typhoons
		.836, .838	Earthquakes
		.834, .794	Transportation accidents
		.850, .771	Tornadoes, strong winds + one special risk for each region
	Large-scale risks	.872, .912	Tsunamis
		.892, .870	Volcanic eruptions
		.860, .886	War, disputes, terrorism
		.879, .875	Accidents at nuclear power facilities
		.897, .795	Chemical pollution, accidents
Vulnerabilities	Topographical vulnerability	.784, .842	Presence of coastlines subject to erosion
		.841, .825	Presence of rapidly flowing rivers
		.699, .783	Presence of low land or land at 0 m elevation
		.862, .738	Presence of active volcanoes
		.834, .600	
	Governmental vulnerability	.823, .812	Weak ties between government and citizens
		.828, .836	Lack of resources for drafting and promoting policies
		.813, .826	Conservativeness of organizations in government agencies
		.808, .832	Weak ties within communities
		.856, .794	Lack of activities by citizens' groups, nonprofits, etc.
	Societal vulnerability	.814, .797	High percentage of low-income earners, poor employment conditions
		.817, .811	High percentage of elderly population, depopulation
		.815, .825	Lack of tax revenue
		.819, .774	High number of single-person households
		.815, .792	High number of vacant homes
			Frequent relocation of residents into and out of the community (low retention)

(continued)

Table 3.2 (continued)

	Factor	Cronbach's reliability coefficient	Relevant observed variables
Situations to be avoided	Crisis situations	.919, .926	Cut-off of food or lifeline supply
		.918, .914	Cut-off of energy supply
		.917, .924	Cut-off or interruption of transportation and telecommunications functions
		.916, .910	Direct damage to human life
		.929, .916	Long-term worsening of food conditions
	Environmental events	.855, .876	Loss of shores, rural land, green land, etc.
		.872, .844	Long-term negative impact on ecosystems
		.858, .835	Sudden, localized worsening of natural environment
		.850, .838	Increase in global warming
		.867, .847	

*The values of Cronbach's coefficient of reliability for each cell read, in order from the upper left to the lower right: Whole (upper left), Sapporo City, Sendai City, Kawasaki City, Niigata Prefecture (lower left), Aichi Prefecture (upper right), Kyoto City, Kagawa Prefecture, Kitakyushu City, and Kagoshima Prefecture (lower right)

indicating the suitability of the variables comprising each factor. As Cronbach's coefficient of reliability for the five observable variables exhibited insufficient values for disaster risks, the factor analysis results for each region were used to add an additional risk that was of special concern to that region. The special concerns are as follows: all regions: "tornadoes, strong winds"; Sapporo City: "cold waves, blizzards"; Sendai City: "cold waves, blizzards"; Kawasaki City: "heat waves, fierce heat"; Niigata Prefecture: "cold waves, blizzards"; Aichi prefecture: "heat waves, fierce heat"; Kyoto City: "heat waves, fierce heat"; Kagawa Prefecture: "tornadoes, strong winds"; Kitakyushu City: "heat waves, fierce heat"; Kagoshima Prefecture: "mudslides, landslides." Moreover, it should be noted that many Cronbach's coefficients of reliability in Kagoshima Prefecture fell below 0.8, showing a topographical vulnerability of 0.6. This is due to Kagoshima Prefecture being the only region with a volcano, making it more likely to have different levels of awareness compared to the other regions in terms of volcano risk and vulnerability. Even factors with slightly lower scores were used to examine variations in structures by region. Therefore, it is important to consider that some factors that are insufficiently applicable to the model were included.

Additionally, to assess the effects of the above three elements on respondent behavior and their perceptions of local government approaches, the numbers of responses indicating suitability of self-support, community support, or public support with regard to the 16 resilience measures were used as variables and incorporated into the model. Incorporating these into the model indicates that it is not possible to correctly calculate a solution using all three approaches (individual/self-help, cooperative, and government). Accordingly, only self-support and community support variables were used.

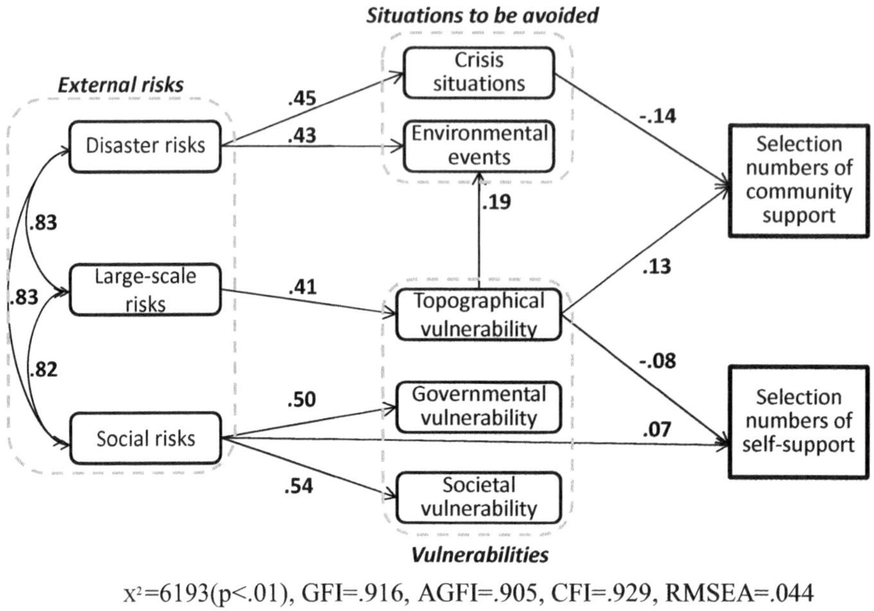

Fig. 3.6 Estimation of the cognitive structure with the civil indicators (whole)

Figure 3.6 shows the results of the covariance structure analysis using the selected variables. The standardized total effects of each factor in relation to the selection frequencies for self-support and community support are shown in Table 3.2. In Fig. 3.6, the error terms were omitted for the sake of clarity, but correlation errors were included for three vulnerability factors and two situations to be avoided. The goodness of fit was sufficient at GFI = 0.916. Significant correlation was observed between risk factors. Thus, respondents are not more aware of specific risks but rather the size of risk perception itself depended on the individual. Meanwhile, the nature of the pertinent risk affected how the respondents assessed vulnerabilities. For example, higher awareness of disaster-related risk influences assessments of situations to be avoided, as does the awareness of large-scale risks on topographical vulnerability and social risks on governmental and societal vulnerabilities. Risks, vulnerabilities, and situations to be avoided were all directly influenced attitudes toward response measures, though some variation in the scope of these effects was observed. Community supports were more likely to be selected as the perceptions of topographical vulnerability increased and were less likely to be selected as the perceptions of crisis situations increased. It is possible that respondents believed that topographical vulnerability should be primarily addressed by the community, whereas this would be difficult in crisis situations where human life is in danger. Preference for self-support measures was lowered by perceptions of topographical vulnerability and increased by perceptions of social risks. It is difficult to respond to a topographical vulnerability as an individual, while respondents may believe themselves capable of self-addressing social risks. Coefficients for self and

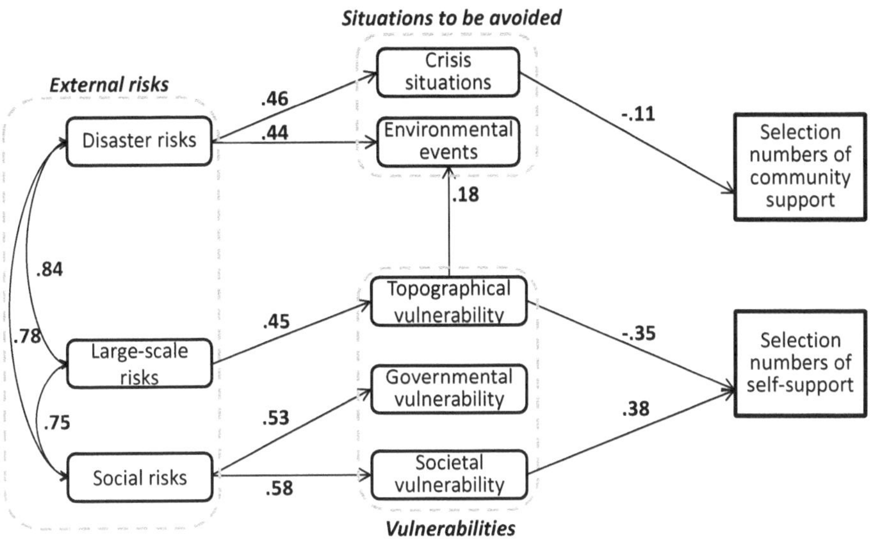

X^2=1692 (p<0.1), GFI = .819, AGFI = .796, CFI = .903, RMSEA = .055

Fig. 3.7 Estimation of the cognitive structure with the civil indicators (Sapporo City)

community support did not exhibit particularly high values. There was a relatively high tendency to prefer public support, with more than half of respondents indicating a preference for public support in 13 out of the 16 relevant measures. This led to the low selection frequency of self and community support.

Figures 3.7, 3.8, 3.9, 3.10, 3.11, 3.12, 3.13, 3.14 and 3.15 show the estimated results by region. It should be noted that the correlation between risk factors for all models was approximately 0.8, the same as that obtained for the all-region model. The pathways drawn from risk to vulnerability factors and factors for situations to be avoided were approximately equal, while significant pathways were drawn from disaster risks to situations to be avoided, from large-scale risks to topographical vulnerability, and from social risks to governmental and societal vulnerability. Therefore, it could be argued that, as a rough structure, commonality exists between each region.

A more detailed inspection showed great divergence between the regions. First, pathways were not drawn from disaster risks to environmental events for Sendai City, Kawasaki City, Niigata Prefecture, Aichi Prefecture, Kyoto City, and Kagawa Prefecture. For these regions, the pathways drawn from social risks showed greater applicability. The pathway drawn from social risks to situations to be avoided for Kyoto City exhibited a particularly higher applicability than those drawn from disaster risks. Thus, social risks are more likely than disaster risks to cause situations to be avoided in Kyoto City. Second, while various methods can be used to draw pathways from vulnerabilities to situations to be avoided, the three regions with significant pathways from governmental vulnerability to crisis situations were Sendai City, Kawasaki City, and Kyoto City. The three regions with significant path-

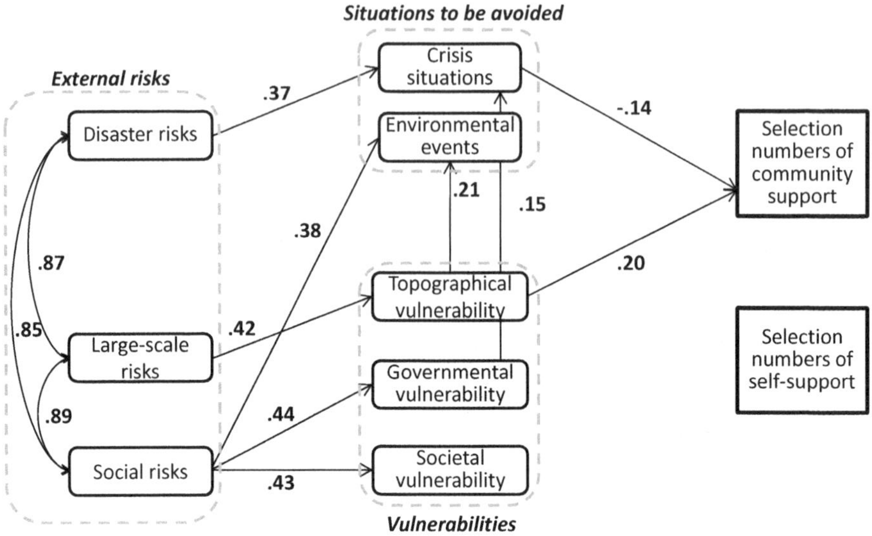

$X^2 = 1693$ ($p<0.1$), GFI = .826, AGFI = .805, CFI = .904, RMSEA = .055

Fig. 3.8 Estimation of the cognitive structure with the civil indicators (Sendai City)

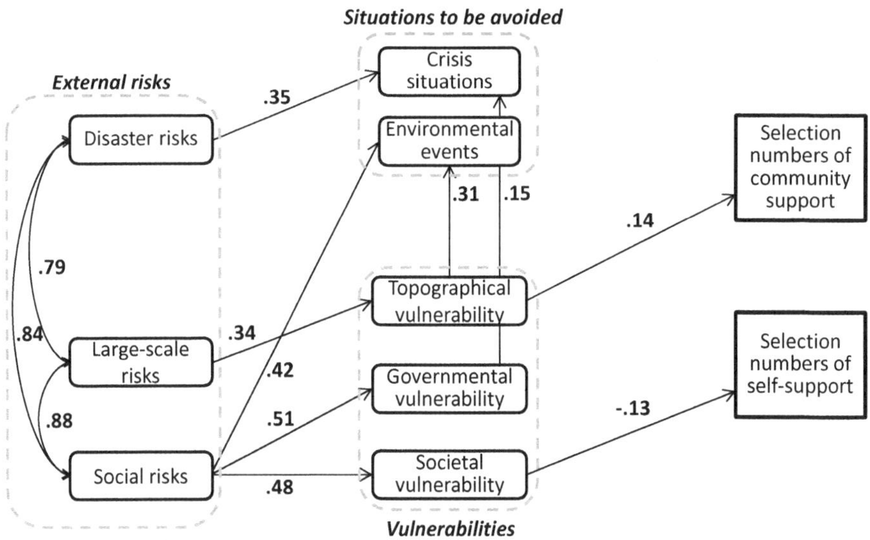

$X^2 = 1693$($p<0.1$), GFI = .835, AGFI = .814, CFI = .905, RMSEA = .053

Fig. 3.9 Estimation of the cognitive structure with the civil indicators (Kawasaki City)

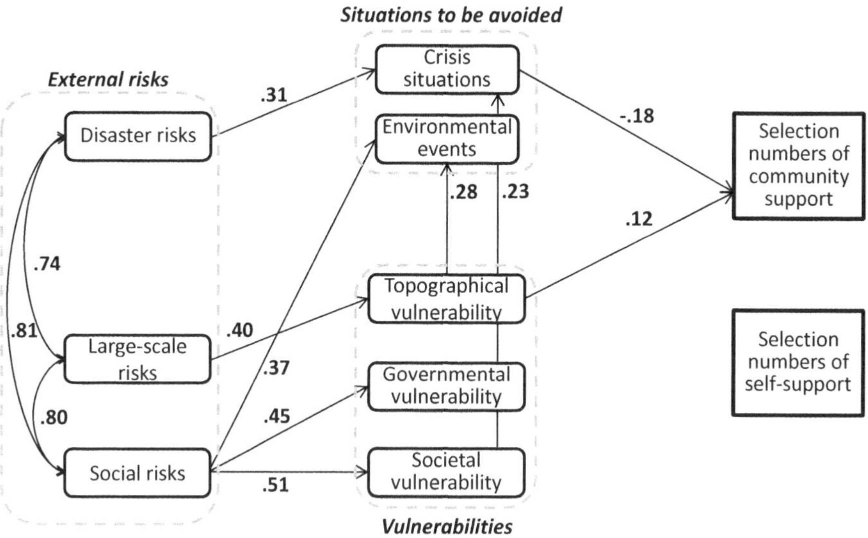

$X^2 = 1632$ ($p<0.1$), GFI = .828, AGFI = .807, CFI = .892, RMSEA = .053

Fig. 3.10 Estimation of the cognitive structure with the civil indicators (Niigata Prefecture)

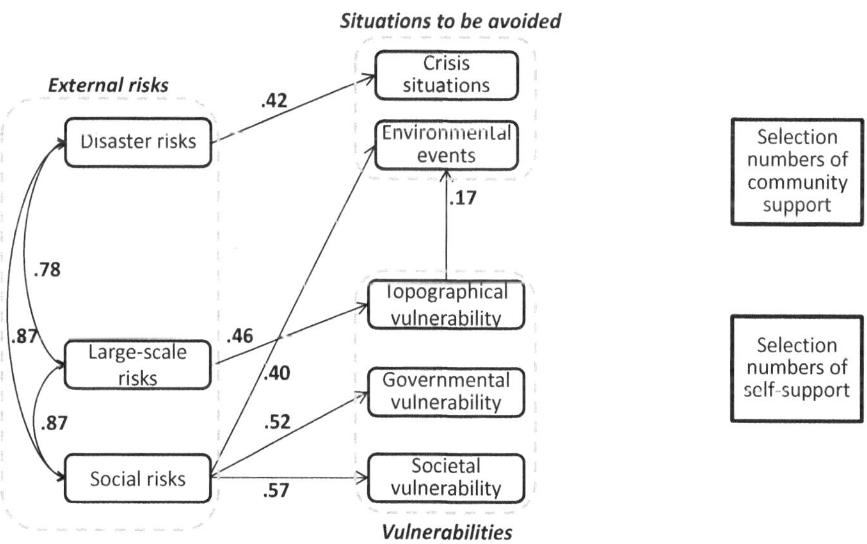

$X^2 = 6791$ ($p<0.1$), GFI = .916, AGFI = .905, CFI = .927, RMSEA = .045

Fig. 3.11 Estimation of the cognitive structure with the civil indicators (Aichi Prefecture)

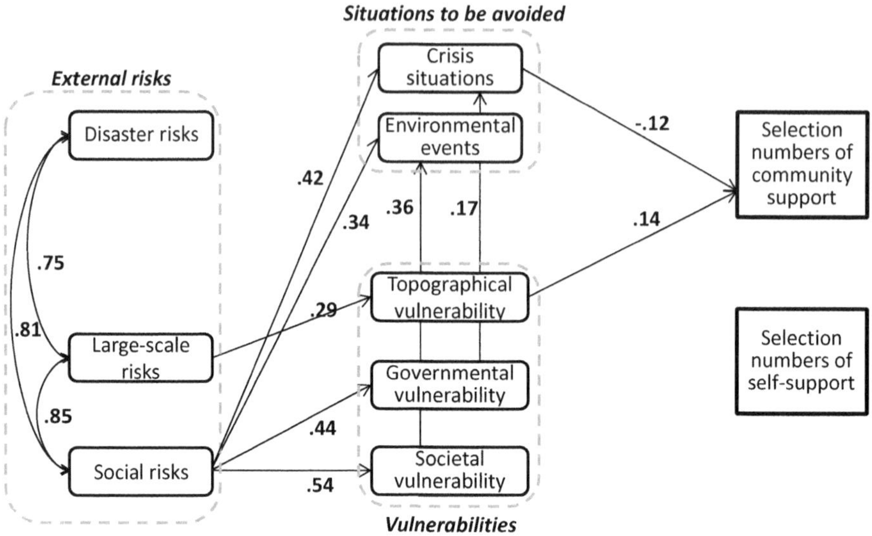

$X^2 = 1702$ ($p<0.1$), GFI = .818, AGFI = .795, CFI = .896, RMSEA = .055

Fig. 3.12 Estimation of the cognitive structure with the civil indicators (Kyoto City)

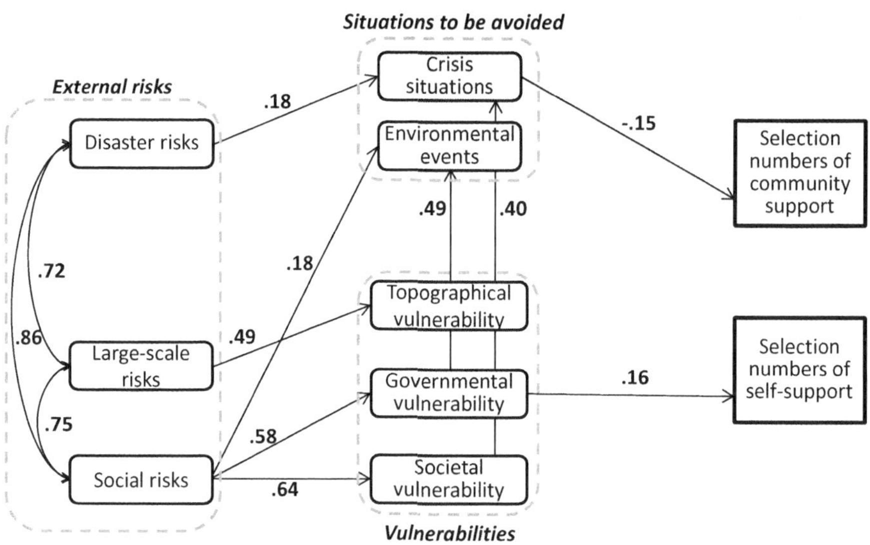

$X^2 = 1608$ ($p<0.1$), GFI = .830, AGFI = .809, CFI = .892, RMSEA = .053

Fig. 3.13 Estimation of the cognitive structure with the civil indicators (Kagawa Prefecture)

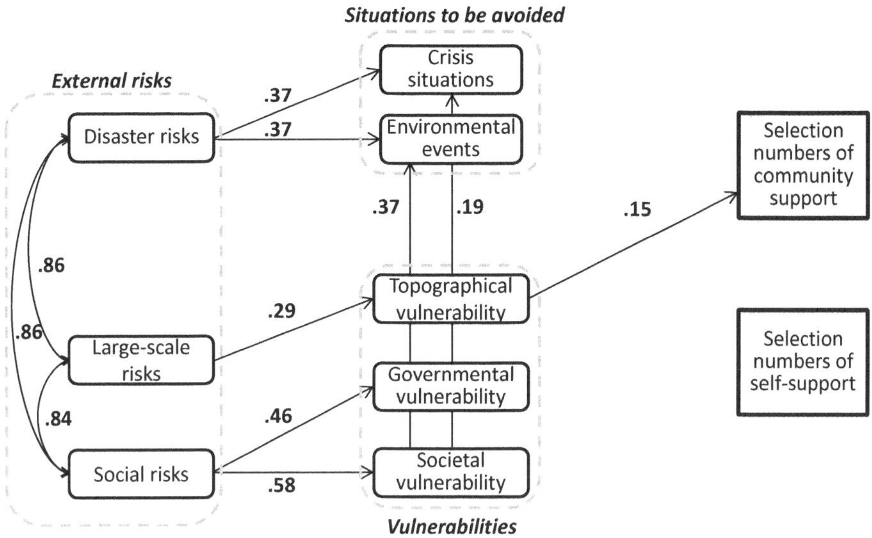

$X^2 = 1733$ ($p<0.1$), GFI = .822, AGFI = .799, CFI = .905, RMSEA = .056

Fig. 3.14 Estimation of the cognitive structure with the civil indicators (Kitakyusyu City)

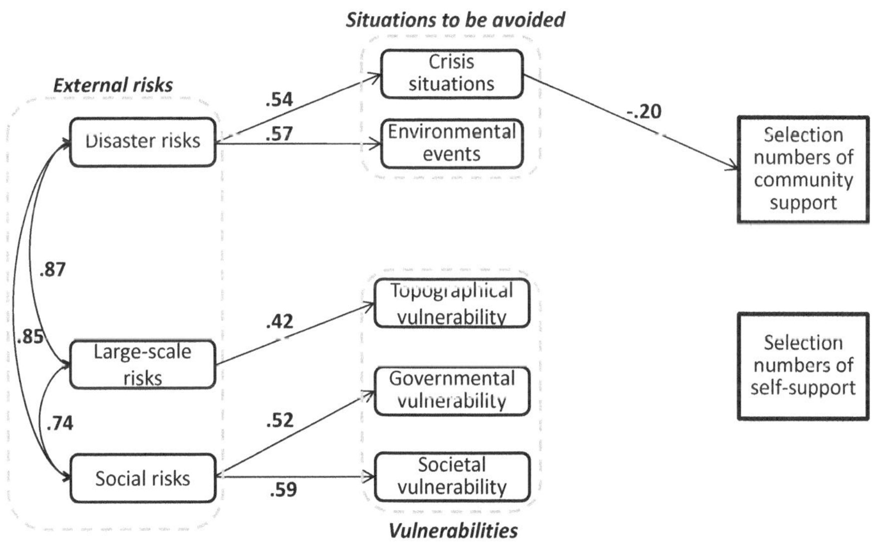

$X^2 = 1857$ ($p<0.1$), GFI = .800, AGFI = .776, CFI = .853, RMSEA = .060

Fig. 3.15 Estimation of the cognitive structure with the civil indicators (Kagoshima Prefecture)

ways from societal vulnerability to crisis situations were Niigata Prefecture, Kagawa Prefecture, and Kitakyushu City. The pathway coefficient to crisis situations for Kagawa Prefecture alone was 0.18 from disaster risks, with a higher value of 0.40 from societal vulnerability, suggesting that serious concerns exist regarding the vulnerability issues of aging and depopulation.

Furthermore, the pathways to selection numbers of self and community support differed significantly by region. The standardized total effects from each factor for self and community support selection are presented in Tables 3.3 and 3.4. First, negative effects from large-scale risks and topographical vulnerability for the all-region results and a positive effect from social risks were observed for self-support. This is likely because topographical vulnerability, such as rivers or lowlands, as well as tsunamis and volcanoes were linked to these concerns. In addition, fears over large-scale risks such as cross-regional incidents were perceived as being difficult to deal with on an individual scale. Moreover, social risks, including population decline and aging, showed increased selections of self-support. This is likely due to an understanding of a greater need to deal with risks on one's own because of concerns regarding social security. Separated by region, only three of the nine regions showed significant pathways: Sapporo City, Kawasaki City, and Kagawa Prefecture. Sapporo City showed high absolute coefficient values, with awareness of each criterion having a significant impact on increasing self-support selection. The number of self-support selections showed slightly higher values in comparison to other regions. Moreover, Kawasaki City exhibited converse results to the all-region study, in that the awareness of social risks and vulnerability decreased self-support selections. This is believed to be related to the fact that Kawasaki City showed significantly lower scores for the following constitutive variables within societal vulnerability: "high number of vacant homes," "high percentage of low-income earners, poor employment conditions," "high percentage of elderly population, depopulation," and "lack of tax revenue."

3.4 Conclusion

With the civil indicators, in terms of a cognitive structure of the general public, we examine the framework of the policy model by analyzing web-based questionnaire data from nine cities in Japan. Major findings are as follows:

First, the highest risk as perceived by the general public is that of "earthquakes," followed by "aging of population," "transportation accidents," "population decreases, low birth rates," and "infectious diseases, viruses." Perceptions of these risks vary significantly with gender, age, and region, suggesting that they accurately reflect the socioeconomic and geographical context of each region.

Second, the vulnerabilities of "high percentage of elderly population, depopulation," "high percentage of low-income earners, poor employment conditions," and "presence of earthquake zones or fault lines" are evaluated as significant. Similar to the aforementioned risk perception, the results of vulnerability evaluations are also significantly different with age and region.

Table 3.3 Standardized total effects on the likelihood to select self-support approaches

	Large-scale risks	Disaster risks	Social risks	Topographical vulnerability	Societal vulnerability	Governmental vulnerability	Crisis situations	Environmental events
Whole	−0.033	0.000	0.066	−0.081	0.000	0.000	0.000	0.000
Sapporo City	−0.160	0.000	0.221	−0.354	0.382	0.000	0.000	0.000
Sendai City	0.000	0.000	0.000	0.000	0.000	0.000	0.000	0.000
Kawasaki City	0.000	0.000	−0.062	0.000	−0.130	0.000	0.000	0.000
Niigata Prefecture	0.000	0.000	0.000	0.000	0.000	0.000	0.000	0.000
Aichi Prefecture	0.000	0.000	0.000	0.000	0.000	0.000	0.000	0.000
Kyoto City	0.000	0.000	0.000	0.000	0.000	0.000	0.000	0.000
Kagawa Prefecture	0.000	0.000	0.096	0.000	0.000	0.165	0.000	0.000
Kitakyusyu City	0.000	0.000	0.000	0.000	0.000	0.000	0.000	0.000
Kagoshima Prefecture	0.000	0.000	0.000	0.000	0.000	0.000	0.000	0.000

Table 3.4 Standardized total effects against community support approaches

	Large-scale risks	Disaster risks	Social risks	Topographical vulnerability	Societal vulnerability	Governmental vulnerability	Crisis situations	Environmental events
Whole	0.055	−0.061	0.000	0.134	0.000	0.000	−0.138	0.000
Sapporo City	0.000	−0.052	0.000	0.000	0.000	0.000	−0.113	0.000
Sendai City	0.084	−0.053	−0.009	0.201	0.000	−0.021	−0.143	0.000
Kawasaki City	0.047	0.000	0.000	0.136	0.000	0.000	0.000	0.000
Niigata Prefecture	0.049	−0.056	−0.022	0.123	−0.043	0.000	−0.185	0.000
Aichi Prefecture	0.000	0.000	0.000	0.000	0.000	0.000	0.000	0.000
Kyoto City	0.039	0.000	−0.060	0.137	0.000	−0.020	−0.121	0.000
Kagawa Prefecture	0.000	−0.027	−0.037	0.000	−0.058	0.000	−0.146	0.000
Kitakyusyu City	0.044	0.000	0.000	0.150	0.000	0.000	0.000	0.000
Kagoshima Prefecture	0.000	−0.108	0.000	0.000	0.000	0.000	−0.200	0.000

Third, "cut-off of food or lifeline supply," "cut-off of energy supply," "long-term worsening of food conditions," and "long-term instability of energy supply" are evaluated as anticipated situations to be avoided. Evaluations of these situations vary significantly with the respondent's level of social capital but do not vary significantly with region. That is, they are typically shared across region.

Fourth, respondents particularly preferred "reviewing and concentration of urban functions" and "moving from high risk areas" as resilience measures by public support. Although more people preferred a self-support approach to "improving energy efficiency and promoting renewable energy" and "moving from high risk areas," for almost all other resilience measures, community support are the second most preferred after public support.

Finally, the cognitive structure whereby perception of risks, evaluations of vulnerability, and situations to be avoided determine attitudes toward resilience measures is generally supported across regions. The nature of the risk affected the object of the respondent's perceptions. For instance, disaster risks influenced assessments of situations to be avoided, large-scale risks influenced perception of topographical vulnerability, and social risks influenced perception of governmental and societal vulnerability. Topographical vulnerability was preferred to be addressed by the local community, which, conversely, may not be able to deal with crisis situations. However, the respondents may consider themselves capable of responding to social risks as individuals.

An important policy implication for local governments is to concentrate on ascertaining and responding to citizen preferences for self or community support to implement resilience measures. Effective responses may include preparing simple guidelines to help citizen carry out the measures they see themselves fit to perform or to improve incentives to encourage them to undertake self-support measures. Conversely, measures aimed at addressing topographical vulnerability that cannot be resolved by individuals, for example long-term, collective relocations, should be performed under government leadership. Policies that are compatible with citizens' real-life experiences and cognitive structures may encourage greater trust and mutual understanding among actors and may lead to further social learning.

Furthermore, the amount of individual variation in risk perception was highlighted. That is, individual differences existed even in relation to risks that were only vaguely perceived. Perception and risk framing are entirely different among the various departments that comprise local governments. Therefore, local governments must respond to citizen perceptions of risk on a day-to-day basis, while also considering the multiple benefits of measures that address problems that are typically unnoticed by citizens, but which are significant over the long term, such as global warming.

Acknowledgment The authors would like to express their sincere gratitude to the respondents who participated in the questionnaire survey and to Ms. Natsumi Ebitani (Graduate school of Agricultural and life sciences, University of Tokyo) and Ms. Izumi Hirata (Graduate school of Engineering, University of Tokyo) who supported for editing.

References

Adger W (2000) Social and ecological resilience: are they related? Prog Hum Geogr 24:347–364
Allen KM (2006) Community-based preparedness and climate adaptation: local capacity building in the Philippines. Disasters 30(1):81–101
Bergamini N, Blasiak R, Eyzaguirre P, Ichikawa K, Mijatovic D, Nakao F, Subramanian SM (2013) Indicators of resilience in socio-ecological production landscapes (SEPLs), UNU-IAS policy report. UNU-IAS. Retrieved August 5, 2015, from https://www.bioversityinternational.org/fileadmin/user_upload/online_library/publications/pdfs/Indicators_of_Resilience_in_Socio-ecological_Production_Landscapes_SEPLs_1676.pdf
Cutter SL, Barnes L, Berry M, Burton C, Evans E, Tate E, Webb J (2008) A place-based model for understanding community resilience to natural disasters. Glob Environ Chang 18:598–606
Godschalk D (2003) Urban hazard mitigation: creating resilient cities. Nat Haz Rev 4:136–143
Hando I, Kubota J (2012) Conceptualism of resilience. In: Kosaka R (ed) Regional resilience: learning from memories of disasters. SHIMIZU KOBUNDO, Tokyo, pp 51–74
ICLEI-Local Government for Sustainability (2012) Preparing for tomorrow strategy 2012–2018, from http://archive.iclei.org/documents/LACS/Publicacoes/BH_Strategic_Plan_final_.pdf
Jha AK, Miner TW, Stanton-Geddes Z (eds) (2013) Building urban resilience: principles, tools, and practice. International Bank for Reconstruction and Development. The World Bank, Washington, DC
Lee MT, Markowitz ME, Howe DP, Ko C, Leiserowitz AA (2015) Predictors of public climate change awareness and risk perception around the world. Nat Clim Chang 5(November):1014–1023
Longstaff PH, Armstrong NJ, Perrin K, Parker WM, Hidek M (2010) Building resilient communities: a preliminary framework for assessment. Homeland Secur Aff 1(3):1–23
National Institute of Environmental Studies of Japan (2014) The report on information of heat stroke patients in 2013 (in Japanese)
Norris FH, Stevens SP, Pfefferbaum B, Wyche KF, Pfefferbaum RL (2008) Community resilience as a metaphor, theory, set of capacities, and strategy for disaster readiness. Am J Community Psychol 41:127–150
Sharma D, Singh R, Singh R (2013) Urban climate resilience: a review of the methodologies adopted under the ACCCRN initiative in Indian cities. Asian cities climate resilience working paper series 5. Retrieved October 14, 2015, from http://pubs.iied.org/pdfs/10650IIED.pdf
The Ministry of Land, Infrastructure, Transport and Tourism of Japan (2012) The situation of drought in Japan in 2012, http://www.mlit.go.jp/river/bousai/main/saigai/kas-sui/pdf/kassui_joukyou_h24.pdf (in Japanese)
Tobin GA (1999) Sustainability and community resilience: the holy grail of hazards planning? Environ Haz 1:13–25
UNISDR. (2012). How to make cities more resilient a handbook for local governments leaders.. UNISDR. Retrieved August 29, 2013, from https://www.unisdr.org/files/26462_handbookfinalonlineversion.pdf
van Aalst MK, Cannon T, Burton I (2008) Community level adaptation to climate change: the potential role of participatory community risk assessment. Glob Environ Chang 18:165–179

Chapter 4
Civil Indicator: The Resilience Index of Regional Communities to the Risks of Disasters

Motoko Kosugi, Kenshi Baba, and Mitsuru Tanaka

Abstract We created a public indicator as an index to measure the resilience of cities and regions against the risk of natural disasters related to climate change, as well as other risks associated with the changing social environment. The public indicator is one of the three main indicators (Baba & Tanaka 2015). In our online survey, we measured the commitment to residential areas, crises that occurred in the regions, and the vulnerabilities of the community. We selected the risks of eight natural hazards, then recruited respondents from eight cities and prefectures that have either high likelihood or a high frequency of experiencing such disasters. Our results indicate that respondents strongly recognized the risk of disasters that frequently occur in their own residential areas as crisis events rather than other disasters. Participation in community activities and a commitment to their localities were high in rural area. The recognition of vulnerability of homes and communities was low in urban areas. These results showed that resilience index of regional community in urban areas such as Sapporo City was high in the aspect of prevention, and resilience index in rural area such as Kagoshima Prefecture was high in the aspect of cooperation and reconstruction.

Keywords Resilience · Regional community · Questionnaire survey · Natural disaster · Climate change

---

M. Kosugi (✉)
Shizuoka University, Shizuoka, Japan
e-mail: kosugi.motoko@shizuoka.ac.jp

K. Baba
Tokyo City University, Yokohama, Kanagawa, Japan
e-mail: kbaba@tcu.ac.jp

M. Tanaka
Hosei University, Machida, Tokyo, Japan
e-mail: mtanaka@hosei.ac.jp

M. Tanaka, K. Baba (eds.), *Resilient Policies in Asian Cities*,
https://doi.org/10.1007/978-981-13-8600-8_4

4.1 Introduction

4.1.1 Background of This Research

It has become clear that the progression of climate change exerts effects in the direction of increasing the scale and frequency of various natural disasters. In the Fifth Assessment Report of the Intergovernmental Panel on Climate Change (IPCC), it is stated that "there is no room to doubt warming," and due to factors such as rising temperatures and sea levels, and reduction of sea ice area in the arctic circle, heat waves, drought, flooding, typhoons, wildfires, and other extreme phenomena will create considerable vulnerability and exposure to ecosystems and humanity (Intergovernmental Panel on Climate Change 2014; Ministry of the Environment 2014). It is clear that climate change is one external force which will have an extremely large impact on the human living environment and social systems, and it will be crucial to improve the sustainability of cities in the face of impacts expected to increase in magnitude going forward.

To boost the resilience of a city, it is crucial to ascertain the current conditions of the city in terms of external force risks, vulnerabilities, and situations to be avoided, and then prepare and implement measures corresponding to those conditions. Therefore, it is necessary to have indicators for measuring resilience. External force risks of cities are factors which cause some damage to the natural and social environment of the city and to the people residing in the region, and it is assumed there is some uncertainty involved in when, and on what scale, such risks will manifest themselves. Baba and Tanaka (2015) proposed three indicators for measuring the resilience of each city: published statistical data (urban indicator), questionnaire surveys of government (administrative indicator), and questionnaire surveys of citizens (civil indicator). For the urban indicator, local government supervisors and experts ascertain/evaluate resilience relating to conditions such as the city's physical infrastructure, economic activity, and environmental factors. For the administrative indicator, local government supervisors and experts check the existence, degree, and progress of past measures, and ascertain/evaluate whether these translate into improved resilience of the city. For the civil indicator, stakeholders, citizens, and experts ascertain/evaluate resilience relating to the conditions of citizen life and environmental factors, in areas such as knowledge/awareness, learning/training, and social capital.

4.1.2 Purpose of This Research

The first purpose of this chapter is to create, based on questionnaire survey data, the "civil indicator" of the three indicators described in previous chapters. As indicated above, the civil indicator is not used alone, and the resilience of the city is ascertained/evaluated in combination with the urban indicator and administrative indicator. Also, it indicates the characteristics of the resilience of the citizens who reside in the pertinent city, and does not imply that standards or objectives exist for evaluating the level of resilience. Thus, by creating civil indicators for multiple cities, the authors indicate the characteristics and levels of citizen resilience of each city in relative terms, and check whether or not they reflect the characteristics of natural disasters and other events experienced at a relatively high frequency by the region subject to the survey.

To raise the resilience of a city, it is crucial to not only conduct scientific evaluation of damage-causing events which can occur there, and make efforts led by national and local government based on the results, but also for each individual living in the regional community to have the defensive capacity, readiness, and adaptability for that. Thus, in this research, the authors look at defensive capacity for external force risks of citizens themselves as an indicator for evaluating resilience of cities from the perspective of the citizen, and measure perception of crisis events and vulnerabilities in the region or community of residence. Also, citizen resilience is exhibited as an entire community, not as the simple sum of resilient individuals (Kulig 1996, 2000; Pfefferbaum et al. 2005; Rose 2004), and therefore elements of social capital such as years of residence and community or neighborhood interaction are included in the indicator. In terms of the ability to recover from damage due to an external force, it has been pointed out that personality traits such as self-confidence and future orientation are important at the individual citizen level (Imamura 2014; Ishihara and Nakamaru 2007; Norris et al. 2008), but a variety of factors are thought to influence individual personality, and their processes of transformation are also diverse. Since the effects on city resilience are not clear, this is not treated in this research.

Furthermore, the authors also gather data on how individuals perceive the various disaster risks to regional society, and examine differences due to the region. Issues such as the order of priority of desirable measures and risk management for disaster risk are likely to be greatly affected by what sort of thing the citizens recognize the pertinent risks to be. By clarifying not only how individual disaster risks are perceived, but also issues such as tendencies and characteristics of perceptions shared by multiple risks, it should be possible to understand the characteristics of resilience of citizens relating to prevention and readiness.

In order to check whether or not the civil indicator composed of the above items differs depending on external force risks of the city or region of residence, respondents are sampled from multiple regions with different frequencies of occurrence of natural disasters. The natural disasters covered here are: landslides, drought, heatstroke, heavy snow falls/cold waves, earthquake/tsunamis.

For more detailed information on the urban indicator, please refer to Kawakubo et al. (2017).

4.2 Procedure

4.2.1 Overview of Survey

The survey was conducted via the Internet on February 9–12, 2015. The survey subjects were 3518 Internet monitors from a survey company (1844 men, 1674 women).

Subjects were sampled from monitors residing in regions with a high frequency of occurrence of natural disasters. The subject areas were: Niigata and Kagoshima Prefecture as regions with a high frequency of occurrence of landslides [over the past 8 years, the two prefectures with the highest frequency of occurrence of debris flows, landslides, and cliff collapse (Water and Disaster Management Bureau, Ministry of Land, Infrastructure, Transport and Tourism 2014)], Kagawa and Aichi Prefecture as regions with a high frequency of occurrence of drought [over the most recent 20 years, two prefectures with a comparatively large number of years in which impacts occurred due to drought (Water and Disaster Management Bureau, Ministry of Land, Infrastructure, Transport and Tourism 2012)], Kyoto city and Kitakyushu city as regions with a high frequency of occurrence of heatstroke [top two cities in terms of occurrence rate of heatstroke patients in fiscal 2013 (National Institute for Environmental Studies 2014)], and Sapporo city as a cold region whose climate differs greatly from most Japanese cities. In addition, for earthquakes the survey focused on Sendai city, which experienced the Great East Japan Earthquake, and is thought to have a high degree of concern about and knowledge of earthquakes and tsunamis.

There was a total of 440 respondents, 400 allocated uniformly in the 8 prefectures/cities in terms of sex and age (5 categories from 20s to 60s and over, in 10-year increments), and 40 government staff sampled beforehand through occupational screening. Relatively speaking, government staff are expected to have greater concern about and knowledge of disaster information and regional vulnerability, i.e., higher resilience, and thus it was felt to be desirable to include them among respondents in order to create the civil indicator.

When there were less than 40 respondents in a cell for 20s or 60s and over, it was decided to ensure the specified number of respondents for each city by, for example, increasing respondents in the neighboring cell (when the cell for men in their 20s was insufficient, the cell for men in their 30s was increased). In addition, due to constraints on the Internet survey technique, it is difficult to sample respondents only from regions with a high frequency of occurrence of disasters (e.g., mountainous areas with many landslides), and thus there is a mixture of cities and prefectures in the subject regions. There are expected to be differences between urban areas and mountainous areas in terms of factors such as direct experience of disasters, but it is likely that a comparatively large amount of information is provided on natural disasters within a prefecture, and thus it was decided that the mixture of cities and prefectures as subject regions would not affect creation of the indicators for resilience.

4.2.2 Composition of Questionnaire

The questionnaire survey administered in order to create the civil indicator was designed to obtain the administrative indicator and civil indicator, and thus many items were inherited from the questionnaire used for creating an administrative indicator on a trial basis in Baba and Tanaka (2015).

The following six types of questions were used to create the civil indicator in this research.

(i) Questions for measuring understanding of crisis events in daily life: Questions were asked on a 5-point scale (1 = Almost no expectation of crisis to 5 = Considerable expectation of crisis) regarding a total of 41 items relating to environmental contamination/pollution such as "noise, vibration" or "air pollution, odor," natural disasters such as "torrential rainfall" or "earthquakes," human-caused environmental destruction such as "decrease in/loss of biodiversity" and "forest depletion," and changes in the social environment such as "population decreases, low birth rates" and "poverty, disparities in living standards." Questions were also asked on a 5-point scale (1 = Completely disagree to 5 = Very much agree) regarding understanding of the impacts and progression of global warming.

(ii) Questions for measuring understanding of vulnerabilities and weak points present in the region of residence or home: Questions were asked on a 5-point scale (1 = Almost not a vulnerability or weak point to 5 = Considerable vulnerability or weak point) regarding a total of 28 items relating to the natural environment such as topography or water resources, urban development such as infrastructure improvement or the quantity of wooden buildings, social environment such as depopulation or weakening of community ties, and government activities such as the conservatism of local government administration or the low level of tax revenue.

(iii) Questions were asked on a 5-point scale (1 = Completely disagree to 5 = Very much agree) regarding 4 items relating to global warming, such as the nature and seriousness of effects.

(iv) Questions for measuring evaluation regarding expert capabilities, reliability, and nature of efforts for organizations such as local government administration and residents' associations in the region of residence.

(v) Questions asking about involvement with regional society, and participation situation and frequency of interaction with neighbors.

(vi) Questions relating to number of years residence at current address, intension to continue residence in the future, and residence status.

(vii) Questions for elucidating perceptions of disaster risk in regional society (described below).

(i) and (ii) are items relating to vulnerability in regional society and the home, and prediction of crises which can occur, and the purpose is to measure the defensive capacity of respondents at the individual level. For (iii), natural disasters which occur at high frequency in the subject regions of this research, such as landslides, intense heat, and cold waves, are included in phenomena whose frequency and scale are likely to increase due to the effects of climate change, and thus the purpose is to ascertain the understanding of that relationship and differences due to the region. (iv)–(vi) are items relating to resilience of the regional community as a whole, such as participation in the regional community and network, and trust in local government administration. Evaluation of and trust in the efforts of government administration are likely to be related to trust in information (sources) and acceptance of warnings/instructions in a crisis, and abundance of social capital and commitment to the region of residence are likely to contribute not only to prevention but also to recovery from disaster damage.

For the perceptions of disaster risk in regional society in (vii), the purpose is to ascertain the characteristics of the cognitive framework, i.e., the priorities in prevention or management, and the type of events/crises the subject disaster risks are understood as. Here, evaluation of individual disaster risks is not important, and the focus is placed on indicating the framework for understanding disasters (disaster risks) which can occur in the region, and ascertaining trends and patterns in how disaster risks are understood in that framework. These trends and patterns in understanding of disaster risks are thought to relate to mindset and readiness toward disaster risks at ordinary times.

The disaster risks for regional society treated in this research were classified based on the characteristics of damage factors and time frames, and eight types were selected for evaluation (Table 4.1). The set of items for evaluating the characteristics of risk, including those which take this sort of long-term, society level as their characteristic, were created by referring to the questions used in prior research involving mapping and intercomparison of evaluation of various risk events (Slovic 1987; Slovic et al. 1979, 1980, 1985). Slovic et al. took as their subject of evaluation risks of actions, technologies, and events in daily life such as smoking, skiing, spray cans,

Table 4.1 Disaster risks in regional society subject to evaluation

	Natural disaster	Climate change	Change in social situation
Short-term effects	(i) Interruption in supply of food or lifelines		
	(ii) Sudden, localized worsening of the natural environment		
	(iii) Collapse, partial collapse, or damage to buildings		
	(iv) Fragmentation or interruption of transportation/communication functions		
Long-term effects		(v) Worsening of water resource situation	(vii) Drop in level of government services
		(vi) Adverse impacts on ecosystem	(viii) Unstable supply of energy

and guns, divided risk evaluation into a "dread factor" and an "unknown factor," and clarified the relative characteristics of various risks on those two axes. In the subjects of evaluation treated in these studies, there are many short-term risks at the individual level, but in this research the subjects of evaluation are disasters in cities and regional society, and thus inappropriate items such as equality of benefits, novelty, or worldwide catastrophe were excluded, and items were added for the anger, unease, and regret people are likely to feel as an emotional reaction when a disaster occurs. In addition, due to the purpose of creating a civil indicator of city resilience, a total of 17 items were used, adding items such as social importance and urgency of efforts as items at the (regional) society level.

4.3 Results

The following section first indicates tabulated results regarding commitment to the region of residence, involvement in the regional society and community, disasters for which there is strong awareness as crisis events, and perceptions of vulnerability of regional societies and homes, and risks in regional society. Then the characteristics of the regions subject to the survey are clarified, and presented as a general civil indicator showing resilience in terms of various aspects.

4.3.1 Commitment of Respondents to Region of Residence

The number of respondents in each region of residence was as indicated in Table 4.2. Regarding years of residence at the current address, the percentage responding "20 years or more" was highest among all regions of residence except Sendai city, and in some cases exceeded 40% of respondents, such as in Niigata Prefecture (44.1%) and Kagawa Prefecture (44.8%). On the other hand, in large cities such as Sendai (26.1%) and Sapporo (28.9%), the figure was 30% of respondents or less. Similarly, the number of respondents living in a detached house (owned house) was high in Niigata Prefecture (71.8%) and Kagawa Prefecture (69.3%), while the number of respondents living in a housing complex (rental) was high in Sendai city (36.1%). Regarding intention to continue residing in the future (measured on a 5-level scale: 1 = No intention at all to continue residing, 2 = Not much intention, 3 = Cannot say either way, 4 = Some intention, and 5 = Strong intention), in all areas more than 20% of respondents chose "strong intention," and about 35–40% chose "some intention." Thus, differences due to the region of residence, like those mentioned above, were not evident.

When subjects were asked about interaction opportunities and organizations participated in, with regard to 11 types of social interaction (such as acquaintances or family, community groups, and hobby activities) respondents indicating interaction with "relatives" were around 50%, interaction with friends/acquaintances around

Table 4.2 Regions of residence of survey subjects and their attributes

Characteristic	Region of residence	Number of respondents (male/female ratio)	Average age	Years of residence (most common range)	Percentage of owned detached houses	Intention to continue (average)
Landslide	Niigata Prefecture	440 (223/212)	45.3	20 years or more	71.8%	3.59
	Kagoshima Prefecture	440 (224/216)	45.2	20 years or more	58.4%	3.53
Drought	Kagawa Prefecture	440 (225/215)	45.2	20 years or more	69.3%	3.73
	Aichi Prefecture	440 (238/202)	45.2	20 years or more	63.0%	3.67
Heatstroke	Kyoto city	440 (230/210)	45.0	20 years or more	53.2%	3.64
	Kitakyushu city	438 (231/207)	45.9	20 years or more	43.2%	3.64
Cold waves, heavy snow fall	Sapporo city	440 (133/207)	45.4	20 years or more	43.4%	3.69
Earthquakes/ tsunamis	Sendai city	440 (232/208)	45.6	1–5 years	42.3%	3.67

30–40%, and “interaction with neighbors” around 30% in all regions of residence. Within those figures, the percentages of respondents indicating the above three types of interaction were high in Kagoshima Prefecture, Niigata Prefecture, and Kagawa Prefecture. On the other hand, in all regions of residence, there were 10–20% of respondents who selected “no applicable option,” and figures were high in Sapporo city (19.1%) and Aichi Prefecture (18.9%). Analysis of variance was carried out to investigate differences between regions, and as a result, a main effect due to region of residence was evident on the number of types of social interaction opportunities ($F(7, 3510) = 4.31$, $p < 0.001$). The number was high in Kagoshima Prefecture and Kagawa Prefecture, and low in Sapporo city and Aichi Prefecture.

In terms of involvement with regional society, subjects were asked about their participation situation in eight types of activities and interaction. There were no differences in the participation situation due to region of residence. The most common type of interaction indicated by respondents was “greet neighbors” at 60–70%, and this was followed by “talk and discuss with neighbors” at 20–30%, “participation in festivals, bazaars, and events” at 20–25%, and “know that within the region the area is susceptible to disasters (cliffs, lowlands etc.)” and “have participated in activities such as shelter setup drills” at about 10%. Also, 20–25% of respondents selected “none applicable.” When a simple sum was taken of those with experience of eight types of involvement in regional society, differences were evident due to region of residence ($F(7, 3510) = 2.52$, $p < 0.05$). It was shown that participation was high in

Kagoshima Prefecture (m = 1.66), Kagawa Prefecture (1.66), and Niigata Prefecture (1.65), and low in Sapporo city (1.41) and Kitakyushu city (1.42).

Due to the above results, differences are evident due to region of residence of respondents in participation in interaction and activities of regional society, years or residence, and residence status. In Niigata Prefecture, Kagawa Prefecture, and Kagoshima Prefecture, participation in activities of regional society is high, years of residence are long, and the percentage of residence in owned detached houses is high, and thus commitment to regional society can be regarded as relatively high. On the other hand, in urban areas such as Sapporo city and Sendai city, the trend is for years of residence to be short, and participation in activities of regional society to be low.

In terms of evaluation of local government administration in the region of residence, subjects were asked to respond regarding the "local government administration" and "organizations of regional society such as residents' associations." They responded on a 5-point scale (1 = Completely disagree to 5 = Very much agree) concerning "On their own initiative, they are working on measures relating to warming and disaster prevention," "they have the expert knowledge to sound the alarm to residents," and "their policies and efforts can be trusted." As a result of analysis of variance, it was found, regarding local government administration, that there is a difference due to region of residence only for "On their own initiative, they are working on measures relating to warming and disaster prevention" ($F(7, 3510) = 2.06$, $p < 0.05$), and respondents from Kitakyushu city (m = 3.02) gave a significantly higher evaluation than respondents from Aichi Prefecture (2.88), Sapporo city (2.88), Kagoshima Prefecture (2.88), and Kagawa Prefecture (2.88). In terms of other items, for regional society organizations such as residents' associations, the average was within the range 2.6–3.0, and no differences were evident due to region of residence. Therefore, it can be said that, in Kitakyushu city, the evaluation of the region's local government is intermediate, and in the remaining regions, the tendency is for local government efforts and capabilities to not be evaluated very highly.

4.3.2 *Expectations of Crisis Events in Daily Life and Understanding of Vulnerabilities in Regional Society*

There are likely to be differences in understanding of crises in daily life depending on what sort of disasters the region of residence is susceptible to. In this research, subjects were asked, regarding 11 disasters/events of the 41 items asked about in the questionnaire, to indicate on a 5-point scale "Do you expect the following events as a "crisis" in your daily life?" (1 = Almost no expectation as a crisis, 2 = Not very much expectation, 3 = Cannot say either way, 4 = Expect to a certain degree, 5 = Considerable expectation), and the average values are tabulated in Table 4.3. This is thought to reflect the understanding of the degree to which the pertinent

Table 4.3 Events expected as crises in daily life

	Landslide		Drought		Heatstroke		Cold wave	Earthquake
	Niigata	Kagoshima	Kagawa	Aichi	Kyoto	Kitakyushu	Sapporo	Sendai
Landslides and debris flows	3.05	**3.41**	3.08	2.90	2.92	3.24	2.99	3.04
Localized torrential rain	**3.53**	**3.53**	3.37	3.46	3.49	3.51	3.32	3.36
Typhoons	3.41	**3.93**	3.50	3.50	3.44	3.56	3.23	3.33
Floods	**3.36**	3.21	3.10	3.17	2.95	3.12	2.97	2.97
Drought and water source depletion	2.99	3.01	**3.58**	3.09	3.00	3.27	2.89	3.02
Heat waves and intense heat	3.34	3.37	3.34	3.29	**3.44**	3.43	3.04	3.21
Cold waves and heavy snow	3.66	2.53	2.70	2.86	2.96	2.98	**3.73**	3.28
Earthquakes	4.02	3.78	3.95	4.08	3.90	3.58	3.74	**4.10**
Tsunamis	3.18	3.23	3.17	3.20	2.65	3.09	2.92	**3.32**
Volcano eruption	2.77	**3.95**	2.50	2.84	2.71	3.01	2.98	3.12

Bold values indicates cells with the highest average of the "degree expected as crisis" for disasters which have had a major effect depending on the area

crisis event can occur in the respondent's region of residence, and higher evaluation points are regarded as an understanding that the event can occur in daily life. One-factor analysis of variance was performed for the regions, and the highest average was shaded among the crisis events for which the main effect is apparent.

For "drought, depletion of water sources (water resources) ($F(7, 3510) = 18.1$, $p < .001$)" it was found, as a result of a multiple test, that respondents in Kagawa Prefecture had a significantly higher average compared to other regions of residence, and thus there was understanding as a "crisis." Similarly, for "heat waves, fierce heat ($F(7, 3510) = 6.63$, $p < .001$)," the average value for respondents from Kyoto city was highest, and a significant difference was evident between Sapporo city and Sendai city in the results of a multiple test. For "cold waves, blizzards ($F(7, 3510) = 69.5$, $p < .001$)," the average for respondents in Sapporo city was the highest, and there was a significant difference with all other regions of residence except for Niigata Prefecture. For "torrential rainfall ($F(7, 3510) = 2.75$, $p < .01$)," respondents from Niigata Prefecture and Kagoshima Prefecture had the highest percentage with expectation as a crisis. For "typhoons ($F(7, 3510) = 19.2$, $p < .001$)" and "mudslides, landslides ($F(7, 3510) = 9.70$, $p < .001$)," the average was highest for respondents from Kagoshima Prefecture, and for "flooding ($F(7, 3510) = 6.96$, $p < .001$)," the average was highest for respondents from Niigata Prefecture. For "earthquakes ($F(7, 3510) = 14.4$, $p < .001$)" and "tsunamis ($F(7, 3510) = 12.7$, $p < .001$)," the average of crisis expectations was highest for respondents from Sendai city.

These results show that, regarding the natural disasters used as criteria when the subject regions were selected, the respondents residing in those regions understood them as crises. However, it cannot really be said that there is a perfect match between disaster damage given by statistical data and crisis understanding. For example, Aichi Prefecture was selected as a region where drought damage occurs at a high frequency, but the crisis expectation of respondents in Aichi Prefecture is significantly lower than that of respondents in Kagawa Prefecture. For landslides too, Niigata Prefecture and Kagoshima Prefecture were selected due to the frequency of disaster occurrence, but while respondents in Niigata Prefecture have a high crisis expectation for localized torrential rain and flooding, their crisis expectation for landslides per se is not high. Both flood damage and landslides are regarded as natural disasters produced by heavy rainfall depending on the features of the topography, and the difference in events expected as crises of respondents from Niigata Prefecture and Kagoshima Prefecture may derive from differences in the topography environment.

At the stage of questionnaire design, this was set as a crisis event for a specific region of residence, but for "volcanic eruptions ($F(7, 3510) = 55.6$, $p < .001$)" respondents from Kagoshima Prefecture had the highest percentage expecting this as a crisis.

With regard to global warming, no differences were seen due to region of residence in perception of its progression or risk ("It is a fact that global warming is occurring ($F(7, 3510)=1.14$, ns.)," "Humanity is being affected by global warming ($F(7, 3510)=.76$, ns.)," "Global warming is a critical problem for myself ($F(7, 3510)=.99$, ns.)," and "Global warming tends to make natural disasters more likely and greater in scale ($F(7, 3510)=.51$, ns.)"). In average values pooling all respondents, understanding that global warming is occurring is the highest ($m = 3.97$), and there is also understanding that it affects natural disasters (3.85), but the understanding that it is a crisis for the subjects themselves was relatively low (3.60).

With regard to vulnerability in the region or home, subjects were asked using a 5-point scale (1 = Almost not a vulnerability or weak point, 2 = Not that great of a vulnerability or weak point, 3 = Cannot say either way, 4 = Certain degree of a vulnerability or weak point, and 5 = Considerable vulnerability or weak point) about a total 28 items: items relating to the natural environmental such as topography or water resources, items relating to urban development such as infrastructure improvement or the quantity of wooden buildings, items relating to the social environment such as depopulation or weakening of community ties, and government activities such as conservatism of local government administration or low level of tax revenue. The subjects of vulnerability evaluation were not just those directly related to the spread of damage in a natural disaster, such as "presence of low land or land at 0 m elevation," "presence of earthquake zones or fault lines," or a "high number or density of wooden homes" but also those relating to prior readiness or recovery after a crisis such as "lack of medical services," or "lack of tax revenue."

Here, if the average evaluation for each item is larger than 3 points (the middle point of the scale), this is regarded as the respondents in that region of residence "understanding that there is a vulnerability." As understanding of respondents, the place where vulnerability was indicated most often was Kagoshima Prefecture (20 items), followed by Kagawa Prefecture (17) and Niigata Prefecture (16). In contrast, the number of items understood as posing a vulnerability was low in Aichi Prefecture (7), followed by Sapporo City (9) and Sendai City (10).

The following four items were understood to be vulnerabilities by respondents in all regions of residence: "high percentage of low-income earners, poor employment," "high percentage of elderly population, depopulation," "weak ties within communities," and "weak ties between government and citizens." In all cases, the specific content related to the social environment in communities.

4.3.3 Evaluation of Disaster Risk in Regional Society

This section examines what sort of characteristics the framework for evaluation of the eight types of disaster risk indicated in Table 4.1 has in each region where respondents reside.

As mentioned above, in the research of Slovic et al., a relative comparison is carried out by mapping risk events using two factors for risk characteristics (dread factor and unknown factor). In this survey too, factor analysis (primary factor method, varimax rotation) was carried out using questions for 17 items prepared based on questions used by Slovic et al.. Also, for all eight types of regional society risks, no difference was evident due to region of residence, and thus it was decided to pool all respondents, and make that the subject of analysis. Due to the nature of the evaluation items making up each factor, Factor 1 is called "severity of risk" and Factor 2 is called "unfamiliarity of risk." In the results of Slovic et al., "dread" is obtained as the first factor and "unknown" as the second factor. "Severity" corresponds to the "dread" factor and "unfamiliarity" to the "unknown" factor, and there is considerable overlap in the items making up each factor. However, the content of the items used in the questionnaire differs somewhat from Slovic et al., and this research includes evaluation from the perspective of the social level, such as social importance and urgency of response, and emotional evaluations such as anger and worry, and thus corresponding names are attached here based on the nature of the items making up the factor.

Taking Factor 1 (severity of risk) as the y-axis and Factor 2 (unfamiliarity of risk) as the x-axis, the average factor loading for each factor was mapped for each region of residence with regard to the eight types of disaster risks. The results are shown in Fig. 4.1.

Taking an overview, the first thing one notices is: risks treated as natural disasters, climate change, and changes in the social environment are not necessarily located close to each other on the map, and that for natural disasters which occur at a high frequency such as landslides or droughts, perceptions of the eight types of

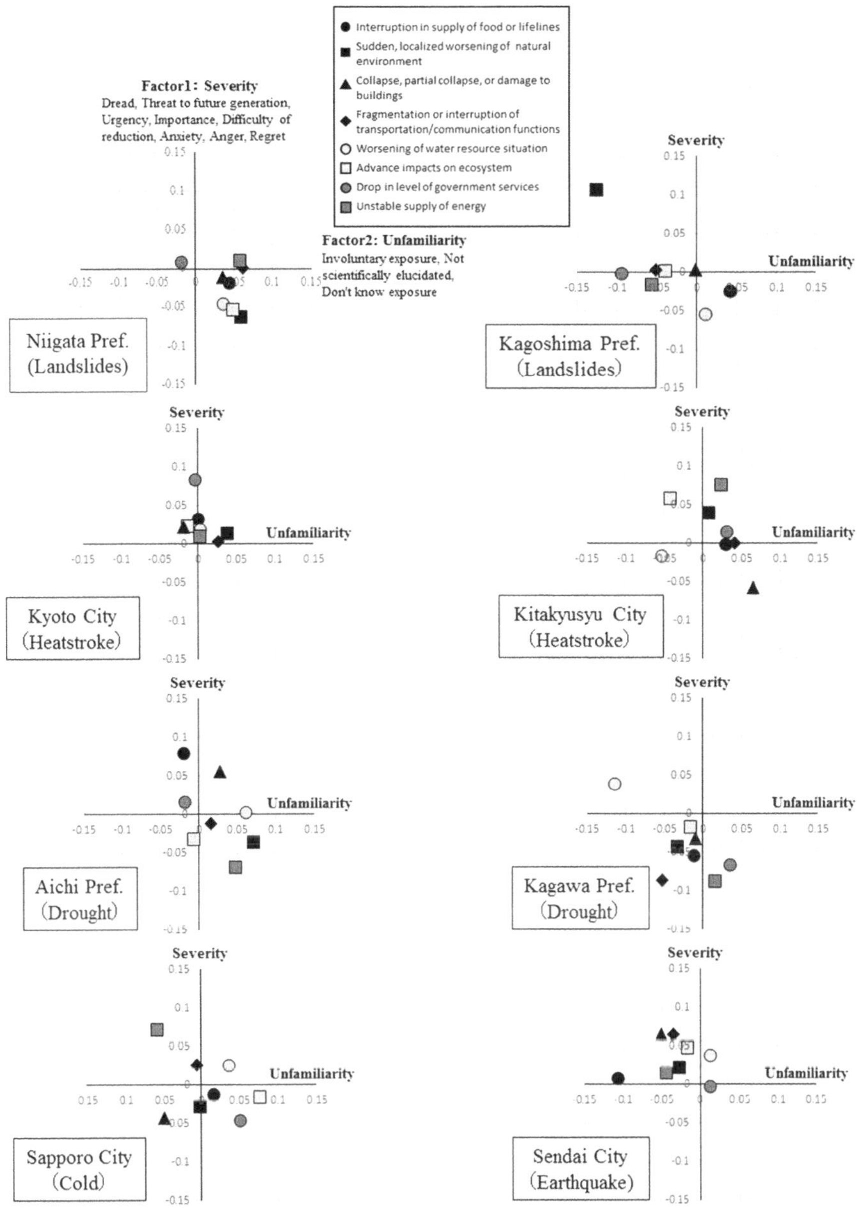

Fig. 4.1 Maps of perceptions of disaster risk in each region of residence

risk taken to be evaluated differ even in the same region of residence. That is, the results suggest that perceptions of risk have different characteristics in each region of residence.

As characteristics due to the region of residence of the respondent, most of the disaster risks are gathered in the second quadrant where risk is severe and familiar for respondents from Sendai city, while many disaster risks are mapped into the fourth quadrant where risk is not severe and unfamiliar for respondents from Niigata Prefecture. For respondents from Kagoshima Prefecture, "a sudden, localized worsening of the natural environment due to a natural disaster (e.g., a landslide or lava flow)" is perceived as a severe, familiar risk, but the other seven disaster risks exhibit little difference in severity, and there is variation in unfamiliarity. For respondents from Kagawa Prefecture, only "worsening of water resource situation due to climate change" was understood as something relatively severe and familiar, but for the remaining disaster risks, there were not very large variations in either severity or unfamiliarity. For respondents from Kyoto city, a "drop in the level of government services due to changes in the social environment" is regarded as a comparatively severe risk. Aside from that, severity and unfamiliarity are gathered around the origin, and variation is small.

4.3.4 Overall Indicators for Each Region

Table 4.4 summarizes characteristics of each civil indicator tabulated thus far for each region of residence of respondents.

"Regional commitment" means years of residence, percentage of owned detached houses, and so forth, and "Involvement with regional society" means the degree of interaction with neighbors and participation in events. While these are strong (or high) in Kagawa Prefecture, Niigata Prefecture, and Kagoshima Prefecture, they are low in Sapporo city, Sendai city, and Aichi Prefecture. However, "Evaluation of local government organization" is an evaluation of efforts and capabilities regarding global warming and disaster prevention of government and resident's association organizations in the region of residence, but there are almost no regional differences, and in terms of evaluation points too, it is lower than the intermediate score (3 = Cannot say either way) of the 5-level evaluation, and thus the overall evaluation of local government organization tended to be low. These are items relating primarily to resilience of the regional community as a whole, and a high evaluation can be regarded as indicating a high degree of mutual aid and recovery capacity of the entire region in case of a disaster.

There are significant regional differences in evaluation of expectation as a crisis of the 11 types of natural disasters, and "Expected crisis events" refers to disasters with the highest average value. However, in Niigata Prefecture, localized torrential rainfall and floods were expected as crisis events more than landslides. "Vulnerability of region/home" gives the results of selecting from the options items which apply to the living environment of the respondent, and the large number of selections indicates that the respondents are placed in a subjectively vulnerable environment. In Kagoshima Prefecture, Niigata Prefecture, and Kagawa Prefecture, the number of selections is high, and in Aichi Prefecture and Sapporo city, it is low. These two items are thought to be involved primarily in the respondents' ability to prevent disasters at the level of individual people and homes, and a proper understanding of

Table 4.4 Overall civil indicators for each region

Region of residence	Regional commitment	Involvement in regional society	Evaluation of local government organization	Expected crisis events	Vulnerability of region/home	Perception of disaster risk
Niigata Prefecture	High		Somewhat low	Localized torrential rain, floods	16/28	Not severe, unfamiliar
Kagoshima Prefecture		High	Somewhat low	Localized torrential rain, typhoons, landslides, volcano eruptions	20/28	Not very severe, large variation in familiarity
Kyoto city			Somewhat low	Heat waves, intense heat	10/28	Small variation in both severity and unfamiliarity
Kitakyushu city			Intermediate		14/28	
Aichi Prefecture		Low	Somewhat low		7/28	Large variation in severity
Kagawa Prefecture	High	High	Somewhat low	Drought	17/28	Not very severe
Sapporo city	Low	Low	Somewhat low	Cold waves	9/28	
Sendai Prefecture	Low		Somewhat low	Earthquakes/tsunamis	10/28	Severe and familiar

disasters with high risk, and grasp of vulnerabilities in the living environment, suggests high capacity to prevent and sense disasters. At the same time, this evaluation indicates the degree to which the respondent is subjectively living in an environment with risk.

"Perception of disaster risk" was characterized on two axes, "severity" and "unfamiliarity," to indicate the tendency of how the disaster risk is perceived. This is not directly related to the level of resilience. It is provided as an aid to reducing bad habits and biases in thinking to improve regional resilience, by clarifying the characteristics of perception of the various disaster risks.

4.4 Conclusion

4.4.1 Summary and Discussion

This research focused on eight regions with a high frequency of occurrence of specific natural disasters, and an indicator was created to evaluate the resilience to disasters of citizens residing in those regions. By using a questionnaire survey, measurements were made of commitment to the region, involvement in the regional society and community, disasters for which there is a strong awareness as a crisis event, vulnerability of regional society and home, perception of risks to regional society, and other factors. These were tabulated, and presented in a form indicating the characteristics for each region of the resilience of regional society, including a variety of aspects.

There are differences depending on the area in years of residence, residence status, interaction with neighbors, and participation in the activities of regional society. In Kagoshima Prefecture, Kagawa Prefecture, and Niigata Prefecture, there is a high degree of participation in the activities of regional society, long years of residence, a high residence rate in owned detached houses, and a relatively high commitment to regional society. In Sapporo city, Sendai city, and Aichi Prefecture, the tendency was toward low participation in the activities of regional society and short years of residence. No regional differences were evident in evaluation of the efforts or expert capabilities of local government administration or regional society organizations, and the overall trend was toward low evaluations. Due to these results, it is thought that Kagawa Prefecture, Kagoshima Prefecture, and Niigata Prefecture ranked high, and Sapporo city, Sendai city, and Aichi Prefecture ranked comparatively low in terms of latent strengths relating to mutual assistance and restoration/recovery in case of a disaster, such as social capital of the regional community and commitment of residents.

There are also regional differences regarding vulnerabilities in living environments such as regions and homes, and the numbers with understanding are high in Kagoshima Prefecture, Kagawa Prefecture, and Niigata Prefecture, and low in Aichi Prefecture and Sapporo city. The fact that numbers with understanding are high means that respondents are living in a subjectively vulnerable environment, and here that is regarded as low resilience. However, low numbers with understanding may also indicate respondents' low ability to sense vulnerability. With regard to the appropriateness of understanding as vulnerability actually present in reality, a separate comparison will need to be made between statistical data and indicators other than resilience. For disasters with a high frequency of occurrence in each region of residence, such as localized torrential rain, drought, and cold waves, people were shown to have a stronger understanding that these disasters are crisis events, compared to other disasters, and it was shown in all regions that natural disasters with high risk are properly understood. This result means that expected crisis events match with disaster frequency in statistical data and disaster experience, and it can be said that at least one part of the civil indicator reflects the characteristics of resilience of the respondents' region of residence. For the above reason, it is likely that resilience, in terms of understanding or prevention of risk in the living environment of an individual, is relatively high in Aichi Prefecture and Sapporo city, and low in Kagoshima Prefecture, Kagawa Prefecture, and Niigata Prefecture.

Regional differences are evident in the perception of disaster risks to regional society, including not only natural disasters, but also changes in the social situation, and in Sendai city there is a trend of disaster risks being perceived in a framework where disasters are severe and familiar. One possible reason for this is that Sendai city has experienced the earthquakes and tsunamis most expected as crisis events, and has come to see other disaster risks as something severe, for which understanding of risk exposure and scientific elucidation are progressing, and at the same time, as a result of post-disaster response, the vulnerabilities of regional society and homes have decreased in relative terms. In Niigata Prefecture, Kagawa Prefecture, and Kagoshima Prefecture, on the other hand, the results suggest that disaster risks tend to be understood as something not that severe. How the various risks are perceived is thought to be related to one's mental attitude toward risk. For example, if there is a tendency to regard risk as something severe and unfamiliar, then there is a possibility that, compared with regions and individuals which do not perceive risk as severe, it will be easier to link with mindset beforehand and the act of making preparations, and that is likely to contribute to regional resilience. Therefore, it may be possible to interpret this as that Sendai city, which tends to perceive disaster risks as a severe, has higher resilience than regions which do not regard risks as very severe.

With regard to global warming, no regional differences were evident in understanding, and it cannot be said that there are differences due to region of residence in understanding at the more general level of global warming. According to the Fifth Assessment Report (2014) of the IPCC, the progression of climate change is increasing the frequency and scale of natural disasters such as heat waves, droughts, floods, and typhoons, and exerting effects on society and ecosystems. Prior research also indicates that perception of global climate variation has effects in the direction of increasing concern with climate change and perception of risk (Lee et al. 2015), and affects behaviors such as making greater contributions to charities relating to global warming (Li et al. 2011), but based on the data in this survey, it was not possible to obtain results suggesting that a relationship has been established between disasters in regions of residence and climate change.

As a civil indicator of overall regional resilience to disasters, the evaluations of Sapporo city and Aichi Prefecture are high, and the evaluations of Kagoshima Prefecture, Niigata Prefecture, and Kagawa Prefecture are low in terms of aspects relating to prevention of regional society risks. On the other hand, Kagoshima Prefecture and Kagawa Prefecture are high, and Sendai city is low in terms of aspects likely to relate to mutual aid and recovery capacity during a crisis. Recovery capacities are crucial, but if greater stress is placed on defensive capacity to reduce the size of damage which might occur in the first place, then as an overall civil indicator, it can be said that resilience of Aichi Prefecture and Sapporo city tend to be high, and resilience of Kagoshima Prefecture and Kagawa Prefecture tend to be low.

4.4.2 Future Tasks

The civil indicator gauges the subjective evaluations and understanding of citizens, and the crisis events and vulnerabilities for a region do not necessary match with the actual vulnerability of social infrastructure and the natural environment (urban indicator) or the specific efforts and progress situation of government (administrative indicator). In order to raise the resilience of cities, it will be important to not only overcome the weaknesses and bolster the strengths exhibited by each indicator, but also to reduce the discrepancies between these indicators. That is, there will be a need to reduce discrepancies in priorities relating to disaster risk and vulnerabilities of regional society to be addressed by government and residents, and to engage in dialog activities to share understanding of what sort of city or region to strive to be through efforts such as raising resilience and managing disaster risk. Going forward, further analysis will be conducted not only on evaluation of individual indicators, but also on relations and correspondences between indicators, and suggestions will be found for designing dialog and decision-making processes of citizens and governments for improving resilience of regional society.

Acknowledgment This research was supported by the Environment Research and Technology Development fund (1-1304) of the Ministry of the Environment, Japan.

References

Baba K, Tanaka M (2015) A concept of Resilient City and providing assessment indicators. J City Plan Inst Japan 50(1):46–53 (in Japanese)

Imamura F (2014) What is resilience in the context of experience of the great East Japan earthquake? Oper Res 59(8):440–445 (in Japanese)

Intergovernmental Panel on Climate Change (2014) Fifth assessment report

Ishihara Y, Nakamaru S (2007) Regarding resilience: the concept, research history, and outlook. Bulletin of Hiroshima Bunkyo Women's College 42:53–81 (in Japanese)

Kawakubo S, Tanaka M, Baba K (2017) Evaluation of resilience of large cities based on public statistical information. J Environ Sci 30(3):215-224 (in Japanese)

Kulig J (1996) Surviving and thriving in the Crowsnest Pass. University of Lethbridge, Regional Centre for Health Promotion & Community Studies, Lethbridge

Kulig J (2000) Community resiliency: the potential for community health nursing theory development. Public Health Nurs 17:374–385

Lee MT, Markowitz ME, Howe DP, Ko C, Leiserowitz AA (2015) Predictors of public climate change awareness and risk perception around the world. Nat Clim Chang 5(November):1014–1023

Li Y, Johnson JE, Zaval L (2011) Local warming: daily temperature change influences belief in global warming. Psychol Sci 22(4):454–459

Ministry of the Environment (2014) Regarding the fifth assessment report (AR5) of the intergovernmental panel on climate change (IPCC) (in Japanese)

National Institute for Environmental Studies (2014) Heatstroke Patient Information Bulletin, Fiscal 2013 Report (in Japanese)

Norris FH, Stevens SP, Pfefferbaum B, Wyche KF, Pfefferbaum RL (2008) Community resilience as a metaphor, theory, set of capacities, and strategy for disaster readiness. Am J Community Psychol 141:127–150

Pfefferbaum B, Reissman D, Pfefferbaum R, Klomp R, Gruwitch R (2005) Building resilience to mass trauma events. In: Doll L, Bonzo S, Mercy J, Sleet D (eds) Handbook on injury and violence interventions. Kluwer Academic Publishers, New York

Rose A (2004) Defining and measuring economic resilience to disasters. Disaster Prev Manag Int J 13(4):307–314

Slovic P (1987) Perception of risk. Science 236(4799):280–285

Slovic P, Fischhoff B, Lichtenstein S (1979) Rating the risks. Environ Sci Policy Sustain Dev 20(3):14–39

Slovic P, Fischhoff B, Lichtenstein S (1980) Facts and fears: understanding perceived risk. Societal risk assessment, pp 181–216

Slovic P, Fischhoff B, Lichtenstein S (1985) Regulation of risk: a psychological perspective. Regulatory policy and the social sciences, vol 241, pp 242–243

Water and Disaster Management Bureau, Ministry of Land, Infrastructure, Transport and Tourism (2012) Drought situation throughout Japan in 2012 (in Japanese)

Water and Disaster Management Bureau, Ministry of Land, Infrastructure, Transport and Tourism (2014) Landslide occurrence situation by prefecture in recent years (in Japanese)

Chapter 5
Administrative Indicator: Local Officials' Cognitive Structure of Policies for Making Resilient Cities

Kenshi Baba, Kosuke Shirai, and Mitsuru Tanaka

Abstract In order to measure administrative indicators, we conducted a questionnaire survey among local governments across Japan. The results show that risks perceived by most local government officials included earthquakes, population decreases, and increases in greenhouse gas emissions. According to the questionnaire results, resilience measures prepared and implemented include the promotion of renewable energy as a precautionary measure as well as the enhancement of methods to collect and provide disaster-related information as an adaptive measure. There are differences among departments of planning, disaster prevention, and environmental policy. The multiple regression analysis results for each department demonstrate that the effects of drivers and barriers in the policy process are stronger than that of the perception and evaluations. As for environmental policy departments, the presence of other local governments to refer to is important in promoting policy. It is likely that the presence of networks of local governments would be important as an agent to promote policy transfer.

Keywords Questionnaire · Risk perception · Vulnerability assessment · Drivers and barriers · Policy process · Preparation and implementation of measures

K. Baba (✉)
Tokyo City University, Yokohama, Kanagawa, Japan
e-mail: kbaba@tcu.ac.jp

K. Shirai
Mitsubishi Research Institute, Inc., Tokyo, Japan
e-mail: kosuke_shirai@mri.co.jp

M. Tanaka
Hosei University, Machida, Tokyo, Japan
e-mail: mtanaka@hosei.ac.jp

M. Tanaka, K. Baba (eds.), *Resilient Policies in Asian Cities*,
https://doi.org/10.1007/978-981-13-8600-8_5

5.1 Introduction

This chapter introduces the results of an evaluation using administrative indicators from the analysis of data derived from questionnaires answered by local government officials throughout Japan. The analyses are then used to evaluate local government officials' external risk perception, as well as their opinions on vulnerabilities and situations to be avoided. The preparation and implementation of each resilience measure is identified through verifying the validity of the structures hypothesized in Fig. 1.2 in Chap. 1. As mentioned in Chap. 1, the structure was generated from literature survey of diversified previous studies such as Adger (2000), Allen (2006), Bergamini et al. (2013), Cutter et al. (2008), Godschalk (2003), ICLEI-Local Government for Sustainability (2012), Jha et al. (2013), Longstaff et al. (2010), Norris et al. (2008), Sharma et al. (2013), Tobin (1999), UNISDR (2012), and van Aalst et al. (2008). We expect to form a clear picture of the cognitive structure of local officials as senders of risk-related information and to highlight the differences in the structure among departments in local governments. By combining results for urban indicators of Chap. 2 and for civil indicators of Chaps. 3 and 4, the authors hope that further mutual understanding and social learning between factors can be realized through the understanding of discrepancies in framing between citizen cognitive structures (the recipients of risk information) and risk managers, such as experts and local officials (the senders). We will discuss the example of Sendai in Chap. 6.

5.2 Methodology

Using the administrative indicators, we designed a questionnaire (summarized in Table 5.1) based on the vast relevant previous studies, in which local government officials were asked to evaluate these indicators. Table 5.2 shows the numbers of responses from different local government types and departments.

Table 5.1 Outline of the questionnaire among local governments

Survey period	March 17–April 25, 2014
Survey subjects	Local officials in planning, disaster prevention, and environmental policy departments of 109 prefectures, ordinance-designated major cities, and mid-sized cities
Method	Sent and collected by post
Questions	Risks from natural and social external forces perceived (41 indicators); vulnerabilities within local communities and government offices (28 indicators); situations to be avoided (24 indicators); the state of preparation and implementation of resilience measures addressing these risks, vulnerabilities, and situations (44 indicators); drivers/barrier of the policy process; attributes of the local government etc.
Responses	Out of 327 distributed questionnaires, 148 questionnaires were answered, accounting for 45.3%

Table 5.2 Responses by local government types and departments

	Planning	Disaster prevention	Environmental policy	Total
Prefectures	17	16	26	59
Ordinance-designated major cities	6	6	10	22
Mid-sized cities	20	18	29	67
Total	43	40	65	148

Questions were chosen to identify perceived natural and social/external risks (41 indicators), vulnerabilities within local communities and government offices (28 indicators), and situations to be avoided (24 indicators). They were the same as those used in the citizen survey that aimed to measure civil indicators discussed in Chaps. 3 and 4. In the questionnaire discussed in this chapter, respondents also were asked to evaluate the state of preparation and implementation of 44 measures to make a city resilient, which were roughly divided into three types: precautionary measures, adaptive measures, and transformative measures. Moreover, we added questions about drivers/barriers of the policy process, attributes of the local government, based on Baba (2005), Baba et al. (2012), and Baba et al. (2017).

We identified that three departments – planning, disaster prevention, and environmental policy departments – are especially involved in the issue of resilience. Therefore, the questionnaires were distributed to the local officials in these departments of 109 prefectures, ordinance-designated major cities, and mid-sized cities throughout Japan. Though the total response rate was not very high, more responses were received from environmental policy departments than from the other two departments. We will basically pay attention to the differences among the departments.

5.3 Results and Observations

First, the analysis method employed here was simply tabulating the evaluations of the 41 indicators concerning risks from external forces, 28 indicators concerning vulnerabilities, and 24 indicators concerning situations to be avoided by each department. In the tabulation, all responses were rated on a five-point scale. We also provided 44 indicators concerning the state of preparation and implementation of resilience measures employing two-valued variables with multiple answers. Then, we examined cognitive structures using covariance structure analysis to verify the structures hypothesized in Fig. 1.2 in Chap. 1 in terms of local officials.

5.3.1 Perceptions of Risk

In Fig. 5.1, significant differences are observed among departments for most risks, with the exceptions of "drought, depletion of water sources (water resources)," "heat waves, fierce heat," "red tides," and "harmful insects." Planning departments showed high scores on long-term situations in local communities such as population

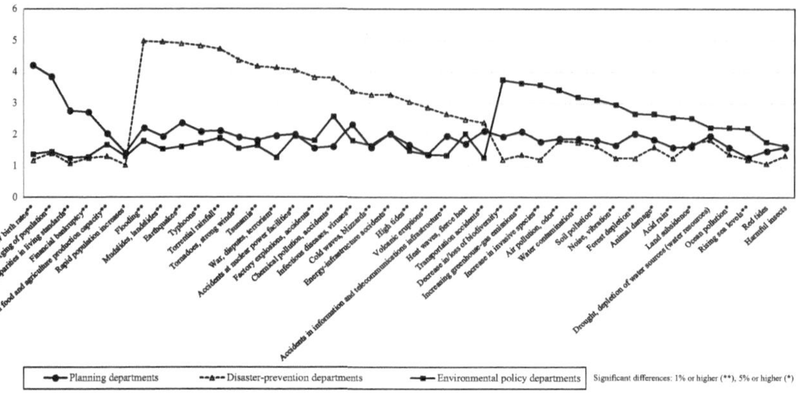

Fig. 5.1 Risk perception caused by external forces (risk/stress)

decreases and aging; disaster prevention departments showed high scores on not just natural external forces but also situations such as human-caused accidents and wars; and environmental policy departments showed high scores on situations related to typical types of pollution as well as long-term situations such as biodiversity and greenhouse gases.

5.3.2 *Assessment of Vulnerabilities*

In Fig. 5.2, among all 28 indicators concerning vulnerabilities within local communities and government offices, significant differences among departments are observed on 13 indicators, such as "presence of steep terrain," "presence of rare or endangered species," and "high percentage of elderly population, depopulation," accounting for less than one-half of the number of indicators prepared. Unlike in the cases of risk perceptions discussed above, many of these indicators are common across local governments and departments, extending beyond the differences in departments' frameworks, which may result in the observed relatively fewer discrepancies among different departments.

5.3.3 *Evaluation of Anticipated Situations to Be Avoided*

In Fig. 5.3, significant differences are observed among departments for most situations, with the exceptions of "long-term worsening of water resources," "worsening of living environment," and "long-term instability in energy supply." Planning departments again showed high scores on long-term situations in local communities

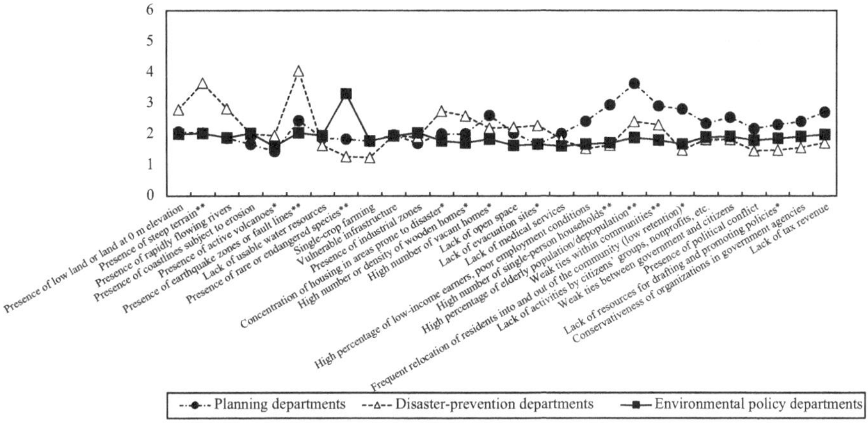

Significant differences: 1% or higher (**), 5% or higher (*)

Fig. 5.2 Vulnerabilities deemed within local communities and government offices

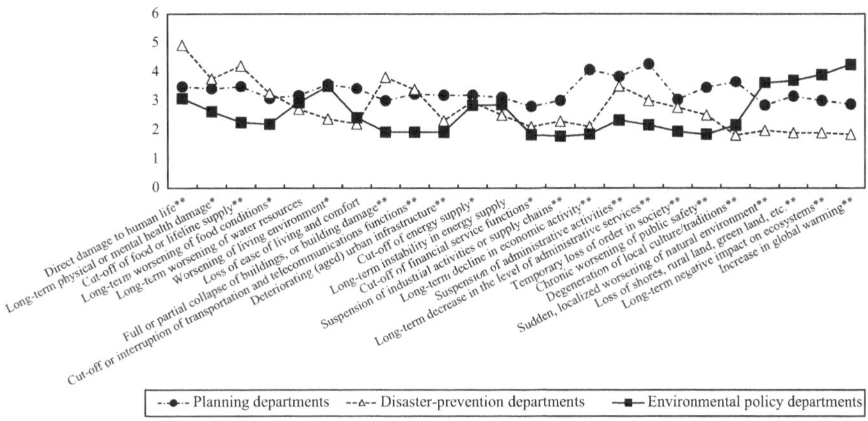

Significant differences: 1% or higher (**), 5% or higher (*)

Fig. 5.3 Situations to be avoided deemed to require resilience measures

such as "long-term decline in economic activity," "long-term decrease in the level of administrative services," "chronic worsening of public safety," and "degeneration of local culture/traditions"; disaster prevention departments showed high scores on severe circumstances in a disaster including "direct damage to human life"; and environmental policy departments showed the highest scores on "increase in global warming." These results strongly reflect the policy framework (the division of responsibilities) of each department.

5.3.4 *Evaluation of Preparation and Implementation of Resilience Measures*

Table 5.3 presents our findings on resilience measures that could be used to respond to the aforementioned risks, vulnerabilities, and situations. We closely examined related plans such as comprehensive plans, community disaster prevention plans,

Table 5.3 The state of preparation and implementation of resilience measures in each department (44 indicators)

	Precautionary measures	Adaptive measures	Transformative measures
Planning departments (N = 17)	Accumulation of government data and coordination with policies (58.8)	Quickly shifting administrative organizations to an emergency footing (47.1)	Centralization of urban functions (realizing a compact city) (29.4)
	Promotion of renewable energy (52.9)	Enhancing methods of collecting and providing disaster information (41.2)	Mean (11.2)
	Mean (37.5)	Mean (35.7)	
Disaster prevention departments (N = 39)	Formation of disaster agreements (100.0)	Quickly shifting administrative organizations to an emergency footing (94.9)	Relocating residents from high-risk areas (10.3)
	Stimulation of self-support, community support, and community functions (92.3)	Enhancing methods of collecting and providing disaster information (89.7)	Mean (2.8)
	Disaster prevention training and public awareness (76.9)	Mean (52.4)	
	Mean (39.7)		
Environmental policy departments (N = 58)	Promotion of energy conservation (96.6)	Enhancing methods of collecting and providing disaster information (32.8)	Development/support of next-generation energy infrastructure (27.6)
	Promotion of renewable energy (93.1)	Quickly shifting administrative organizations to an emergency footing (24.1)	Mean (7.1)
	Nature conservation & promotion of forestation (72.4)	Mean (11.5)	
	Mean (26.1)		

Note: Figures in parentheses indicate the percentage of departments implementing or preparing such measures. Measures in each cell are ordered by rank from the first to the third

and global-warming prevention plans in the cities of Sendai, Kawasaki, Nagoya, and Toyota, where we conducted the interview. Based on the results of this examination, we sought out these 44 indicators by preparing 19 indicators for precautionary measures, 14 for adaptive measures, and 11 for transformative measures. In addition, we employed two-valued variables (with multiple answers) for the evaluation of the state of resilience measures preparation and implementation.

Three overall tendencies can be identified for preparation and implementation of resilience measures: (i) the rates of transformative measures preparation and implementation are markedly low. Among these measures, the most frequently cited one was "development/support of next-generation energy infrastructure," followed by "centralization of urban functions (realizing a compact city)." (ii) Although all departments reported implementing similar adaptative measures, the percentage of disaster prevention departments citing such measures was much higher than those of environmental policy departments and planning departments. (iii) Differences in frameworks were apparent among departments in terms of precautionary measures: 100% of disaster prevention departments reported the "formation of disaster agreements" and nearly 100% of environmental policy departments reported that they were "promotion of energy conservation."

5.3.5 Examining the Cognitive Structure

Based on the results from above analyses, using the same method that we verified the cognitive structure of civil indicators in the preceding chapter, we conducted a similar model analysis for administrative indicators using covariance structure analysis. The sample size decreased to 106 as a result of eliminating answers with missing values. Figure 5.4 shows the results of the model estimation using administrative indicators, while Fig. 5.5 reprints the results of the model estimation using civil indicators for reference purpose.

Comparison with the results using civil indicators shows that there are variations in correlations among risk factors: While there is a very high correlation between disaster risks and large-scale risks, the correlations between large-scale risks and social risks and between disaster risks and social risks are not very strong, at 0.48 and 0.29, respectively. Based on these results, it can be concluded that a gap in perception arises from the fact that while citizens have a vague sense of varying degrees of unease about a wide range of various risks, without focusing on the properties of individual risks, administrators are more sensitive to individual risk properties. While the significant paths are largely the same as in the case of civil indicators, for nearly all paths the coefficients are higher than those in the case of civil indicators. That is, the effects on perception between individual factors can be stronger for local officials than for citizens. Moreover, the sign of the coefficient for the path from disaster risks to environmental events is negative – the opposite of that in the case of civil indicators. Different from disaster risks that must be responded to immediately when an incident arises, environmental events have effects becoming

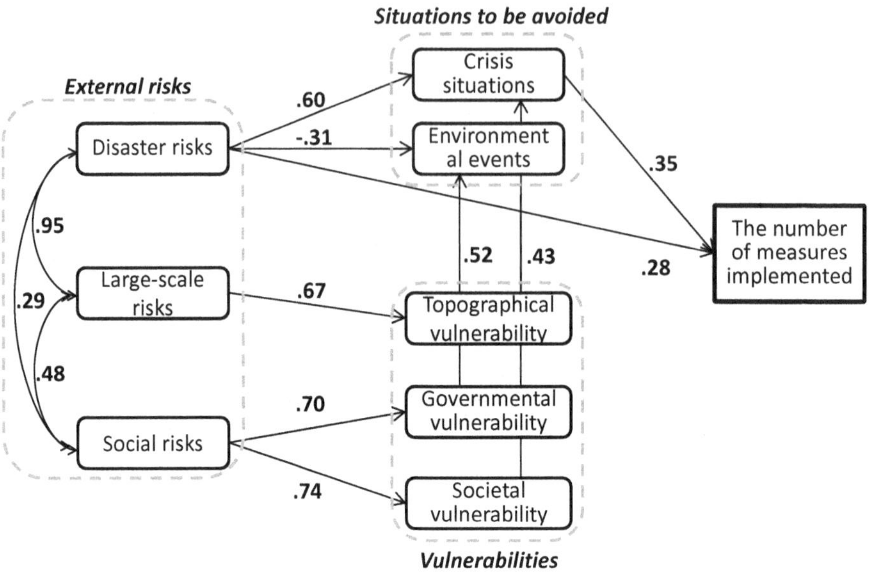

X^2 = 1577 (p<.01), GFI =.592, AGFI =.538, CFI =.838, RMSEA =.106

Fig. 5.4 Estimation of the cognitive structure with the administrative indicators

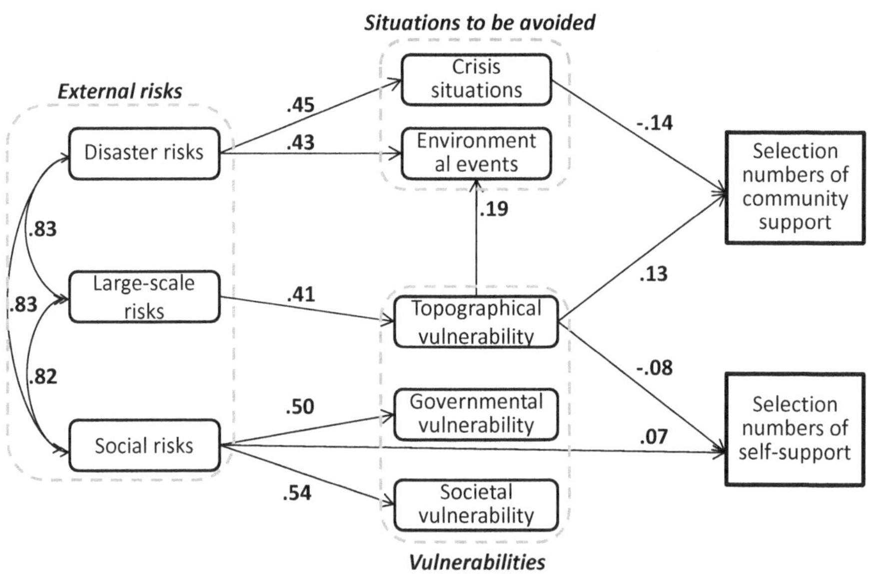

X^2 = 6193 (p<.01), GFI =.916, AGFI =.905, CFI =.929, RMSEA =.044

Fig. 5.5 Estimation of the cognitive structure with the civil indicators (reprint for reference)

Table 5.4 Number of resilience measures prepared and implemented

	Precautionary measures	Adaptive measures	Transformative measures	Total
Planning	7.05	4.68	1.16	12.89
Disaster prevention	7.70	7.38	0.35	15.43
Environmental policy	4.95	1.67	0.78	7.40
Prefectures	6.33	3.79	0.72	10.84
Ordinance-designated major cities	6.10	3.60	0.70	10.40
Mid-sized cities	6.20	4.56	0.67	11.43
Total	6.23	4.11	0.69	11.03

apparent over a long time span. Since the local governments especially disaster prevention departments bear active responsibility for responding to incidents, this difference between disaster risks and environmental events can be considered as a distinguishing feature of administrative indicators.

In addition, in the administrative indicator model, the number of resilience measures prepared and implemented by local governments is used as a variable. In comparison, as discussed in the preceding chapter, in the civil indicators model, the number of resilience measures selected as community support or self-support is used as a variable for citizens' acceptability to resilience measures. A significant path is drawn from disaster risks to crisis situations in the model. This may result from the fact that while disaster risks mainly are under the jurisdiction of disaster prevention departments, disaster prevention departments implement more than twice as many measures as environmental policy departments do (Table 5.4).

5.3.6 *Exploring Other Cognitive Structures in Each Department*

5.3.6.1 Factor Analysis Conducted for Risk Perception, Evaluation of Vulnerabilities and Situations to Be Avoided, and the Number of Resilience Measures Prepared and Implemented

As discussed before, although the model was prepared in the same way as for civil indicators, the goodness of fit for administrative indicators was low. One possible reason for this is the small sample size (148 responses in total, Table 5.2); another reason is the large differences among departments in the perception of risks and situations to be avoided. Hence, ideally, an analysis should be conducted separately for each department. However, due to the limited sample size, it would be difficult to employ covariance structure analysis. Instead, we conducted a multiple regression analysis with the number of measures implemented as the response variable.

First, we conducted a factor analysis (method of maximum likelihood with promax rotation) for each factor of the perception of risks of external forces, evaluation of vulnerabilities, and evaluation of situations to be avoided. Table 5.5 shows the results for the perception of risks of external forces. According to the variables making up each of the factors in terms of the scores by department, for the first factor (f1), disaster prevention departments had the highest scores on all items; for the second factor (f2), environmental policy departments had the highest scores on all items except harmful insects; and for the third factor (f3), planning departments had the highest scores on all items. As a result, we named f1, f2, and f3 "disaster risk," "environmental risk," and "planning risk," respectively.

Next, Table 5.6 shows the results of factor analyses conducted for the evaluation of vulnerability, which are different from that for the risk perception of external forces: for the first factor, planning departments had the highest scores on all items; for the second factor, disaster prevention departments had the highest scores on all items except "vulnerable infrastructure"; and for the third factor, environmental policy departments had the highest scores on all items. Thus, we named f1, f3, and f3 "planning vulnerabilities," "disaster vulnerabilities," and "environmental vulnerabilities," respectively.

Furthermore, Table 5.7 shows the results of factor analyses conducted for the evaluation of situations to be avoided: for the first factor, disaster prevention departments had the highest scores on all items with the exceptions of "cut-off of energy supply" and "long-term worsening of water resources"; for the second factor, planning departments had the highest scores on all items; and for the third factor, environmental policy departments had the highest scores on all items except "worsening of the living environment" and "long-term instability in energy supply." Therefore, we named f1, f2, and f3 "disaster situations," "planning situations," and "environmental situations," respectively.

Table 5.8 shows the correlations between each factor and the number of resilience measures implemented. Due in part to an overall large number of measures implemented by disaster prevention departments, the correlation was high between disaster risks, disaster vulnerabilities, and disaster situations – all factors made up mainly of disaster-related items. Moreover, correlation values were slightly higher for planning factors than that for environmental factors. In addition, the correlation with the number of adaptive measures implemented for environmental factors was particularly low.

Then, we conducted a multiple regression analysis for each department using the number of measures implemented as the response variable and the nine factors above as the explanatory variables. However, only one variable was identified as significant for each department and its R^2 value was not high. In other words, the variables of the risk perception of external forces, evaluation of vulnerabilities, and evaluation of situations to be avoided can be considered as insufficient to explain the number of resilience measures implemented.

Table 5.5 Results of factor analyses of the risk perception of external forces

Risk perception	f1	f2	f3	Risk perception	f1	f2	f3
Mudslides, landslides	**1.064**	−0.172	−0.111	Water contamination	−0.001	**1.054**	−0.235
Flooding	**1.050**	−0.064	−0.126	Noise, vibration	−0.127	**0.981**	−0.063
Typhoons	**1.020**	−0.061	−0.116	Decrease in/loss of biodiversity	−0.239	**0.863**	0.020
Torrential rainfall	**1.011**	0.000	−0.140	Increase in invasive species	−0.173	**0.848**	−0.011
Earthquakes	**1.006**	−0.078	−0.078	Acid rain	−0.025	**0.842**	0.037
Tornadoes, strong winds	**0.983**	−0.106	−0.070	Increasing greenhouse-gas emissions	−0.131	**0.825**	0.013
Tsunamis	**0.883**	−0.032	−0.102	Land subsidence	0.072	**0.817**	0.020
War, disputes, terrorism	**0.857**	−0.268	0.122	Animal damage	0.125	**0.644**	0.065
Accidents at nuclear power facilities	**0.779**	−0.047	0.032	Ocean pollution	0.143	**0.621**	0.108
Volcanic eruptions	**0.736**	−0.100	0.060	Forest depletion	−0.023	**0.584**	0.287
Cold waves, blizzards	**0.730**	0.007	0.056	Rising sea levels	0.080	**0.540**	0.237
Factory explosions, accidents	**0.719**	0.129	−0.085	Drought, depletion of water sources (water resources)	0.216	**0.539**	0.161
High tides	**0.712**	0.027	0.035	Red tides	0.146	**0.497**	**0.310**
Infectious diseases, viruses	**0.634**	0.136	0.145	Harmful insects	0.229	**0.479**	**0.345**
Chemical pollution, accidents	**0.601**	0.310	−0.189	Population decreases, low birth rates	−0.201	−0.140	**1.028**
Energy-infrastructure accidents	**0.575**	0.145	0.066	Aging of population	−0.102	−0.108	**1.004**
Accidents in information and telecommunications infrastructure	**0.540**	0.079	0.325	Financial bankruptcy	−0.066	−0.090	**0.998**
Heat waves, fierce heat	**0.511**	0.320	0.162	Poverty, disparitics in living standards	−0.051	−0.034	**0.980**
Transportation accidents	**0.446**	0.107	0.403	Rapid population increases	0.082	0.132	**0.696**
Air pollution, odor	−0.048	**1.086**	−0.273	Eigenvalue	18.9	8.0	3.1
Soil pollution	−0.027	**1.069**	−0.233	Percentage of cumulative distribution explained (%)	43.2	63.4	70.8

Table 5.6 Results of factor analysis of vulnerability evaluation

Vulnerabilities	f1	f2	f3	Vulnerabilities	f1	f2	f3
Frequent relocation of residents into and out of the community (low retention)	**1.015**	−0.098	−0.081	Presence of low land or land at 0 m elevation	−0.059	**0.961**	−0.071
High number of single-person households	**1.009**	−0.048	−0.162	Concentration of housing in areas prone to disaster	0.043	**0.911**	−0.035
High percentage of elderly population, depopulation	**0.995**	0.090	**−0.354**	High number or density of wooden homes	0.209	**0.821**	−0.079
Weak ties between government and citizens	**0.913**	0.029	−0.004	Presence of rapidly flowing rivers	0.030	**0.778**	0.101
High percentage of low-income earners, poor employment conditions	**0.907**	0.026	0.015	Lack of evacuation sites	0.152	**0.736**	0.049
Weak ties within communities	**0.889**	0.235	**−0.315**	Presence of coastlines subject to erosion	−0.072	**0.677**	0.281
Lack of resources for drafting and promoting policies	**0.886**	−0.160	0.248	Lack of open space	0.292	**0.604**	0.063
Lack of tax revenue	**0.862**	−0.092	0.084	Presence of active volcanoes	0.039	**0.401**	**0.314**
Conservativeness of organizations in government agencies	**0.838**	−0.150	0.223	Vulnerable infrastructure	**0.334**	**0.356**	0.277
Presence of political conflict	**0.815**	−0.112	0.278	Presence of rare or endangered species	−0.213	−0.079	**0.871**
Lack of activities by citizens' groups, nonprofits, etc.	**0.808**	0.076	0.069	Single-crop farming	**0.300**	0.035	**0.690**
Lack of medical services	**0.582**	0.288	0.091	Presence of industrial zones	0.070	**0.365**	**0.435**
High number of vacant homes	**0.560**	**0.380**	−0.091	Lack of usable water resources	0.237	0.269	**0.421**
Presence of steep terrain	−0.088	**1.045**	−0.121	Eigenvalue	18.1	2.7	1.3
Presence of earthquake zones or fault lines	−0.205	**1.013**	−0.079	Percentage of cumulative distribution explained (%)	63.1	72.6	76.0

Table 5.7 Results of factor analyses of situations to be avoided

Situations to be avoided	f1	f2	f3	Situations to be avoided	f1	f2	f3
Direct damage to human life	**1.049**	**−0.339**	−0.059	Temporary loss of order in society	0.290	**0.650**	0.019
Cut-off of food or lifeline supply	**1.000**	−0.079	−0.099	Chronic worsening of public safety	0.259	**0.650**	0.082
Full or partial collapse of buildings, or building damage	**0.872**	0.137	−0.179	Loss of ease of living and comfort	0.086	**0.567**	0.273
Long-term physical or mental health damage	**0.849**	−0.026	0.041	Cut-off of financial service functions	0.277	**0.543**	0.225
Cut-off or interruption of transportation and telecommunications functions	**0.790**	0.206	−0.068	Suspension of industrial activities or supply chains	**0.347**	**0.541**	0.158
Long-term worsening of food conditions	**0.701**	0.193	0.069	Long-term negative impact on ecosystems	−0.099	−0.130	**1.073**
Cut-off of energy supply	**0.496**	0.093	**0.321**	Increase in global warming	−0.049	−0.210	**1.010**
Long-term worsening of water resources	**0.413**	0.204	**0.345**	Loss of shores, rural land, green land, etc.	−0.205	0.067	**0.999**
Long-term decline in economic activity	−0.261	**1.157**	−0.077	Sudden, localized worsening of natural environment	0.101	−0.025	**0.808**
Degeneration of local culture/traditions	**−0.330**	**0.998**	0.175	Worsening of living environment	0.044	0.228	**0.566**
Long-term decrease in the level of administrative services	0.074	**0.937**	−0.292	Long-term instability in energy supply	**0.333**	0.171	**0.462**
Suspension of administrative activities	0.273	**0.714**	−0.214	Eigenvalue	14.7	3.1	1.4
Deteriorating (aged) urban infrastructure	0.255	**0.664**	0.033	Percentage of cumulative distribution explained (%)	59.8	72.3	76.9

Table 5.8 Correlation between perception factors and the number of measures implemented

	Disaster risks	Environmental risks	Planning risks	Planning vulnerabilities	Disaster vulnerabilities	Environmental vulnerabilities	Disaster situations	Planning situations	Environmental situations
No of precautionary measures implemented	0.443**	0.202*	0.259**	0.264**	0.502**	0.186	0.476**	0.359**	0.268**
No of adaptation measures implemented	0.602**	0.064	0.222*	0.231*	0.527**	0.111	0.558**	0.358**	0.077
No of transformative measures implemented	0.104	0.127	0.233*	0.151	0.164	0.201*	0.179	0.170	0.199*
Total No of resilience measures implemented	0.534**	0.154	0.276**	0.270**	0.537**	0.180	0.541**	0.383**	0.205*

* $p < .05$, ** $p < .01$

5.3.6.2 The Relationship Between Policy Formulation Drivers and Barriers, and the Number of Resilience Measures Prepared and Implemented

To consider other factors, we conducted a principal component analysis using a varimax rotation of 15 factors of policy formulation drivers and 15 factors of policy formulation barriers. Since no local governments cited the "presence of previous failures within the administrative organization" as a barrier, it was excluded from this analysis.

As a result, six principal components were identified for drivers and barriers. For drivers, we named f1–f6 as "internal/external proactivity," "participation of other organizations," "presence of other local governments," "securing budgeting," "existing knowledge," and "past experience/desires of residents," respectively (Table 5.9). Similarly, for barriers, we named f1–f6 as "shallowness of external relations," "lowness of external interest," "external opposition," "shortage of knowledge," "lowness of internal interest," and "lack of experience," (Table 5.10).

Table 5.11 shows the single correlation between each factor and the number of resilience measures prepared and implemented in each department. As shown in the table, we were unable to achieve significant results for various variables. However, for drivers, there is a robust tendency for "securing budgeting" and "past experience/desires of residents" to increase numbers of measures. However, no barrier was observed to have a significant negative effect. On the contrary, "shallowness of external relations" and "lowness of external interest" have the positive effect of significantly increasing the number of measures, suggesting that in some cases measures are promoted even in situations such as when subsidies are not secured, there is a lack of previous models, a lack of exchange of information, insufficient discussion takes place, and levels of interest are low among business and industry. There is a need for a closer examination of this matter.

5.3.6.3 Examining Correlative Relationships Among Each Factor in Each Department

In the last section of this chapter, we conducted a multiple regression analysis for each department using the number of resilience measures implemented as the response variable, based on analysis results described in Sects. 5.2 and 5.3. Considering multicollinearity, we extracted only those with the variance inflation factor (VIF) values of 2 or lower. The results are tabulated in Table 5.12.

First, planning departments have a very high level of explanatory power, with an adjusted R^2 of 0.941. As explanatory variables, the factors of the risk perception, and evaluation of vulnerabilities and situations to be avoided are not significant, indicating that the effects of drivers and barriers in the policy process are stronger than those of these perception factors. The drivers of "past experience/desires of residents" and "securing budgeting" have particularly strong effects. In addition, the

Table 5.9 Results of principal component analyses of policy formulation drivers

Policy formulation drivers	f1	f2	f3	f4	f5	f6
Active attitude on part of head, and presence of policies	**0.670**	0.028	−0.039	0.081	−0.020	0.114
Lively discussion	**0.658**	0.002	0.043	0.089	0.293	0.071
Active encouragement by council and its membership	**0.557**	0.278	**0.303**	−0.039	0.118	−0.230
Participation of citizens' groups and nonprofits/NGOs	0.227	**0.763**	−0.158	0.060	0.141	−0.005
Participation of businesses and industry organizations	0.027	**0.760**	0.182	0.154	−0.112	0.096
Accumulation of experience and specialization within the administration	**−0.372**	**0.508**	0.081	0.001	**0.484**	0.100
Presence of prior models by other local governments	−0.012	0.190	**0.850**	−0.122	0.054	−0.019
Exchange of information with national government, prefecture, etc.	0.114	−0.206	**0.758**	0.299	0.121	0.055
Active budgeting measures within the administration	−0.070	0.034	−0.095	**0.855**	0.176	0.129
Securing of subsidies from national government, prefecture, related foundations, etc.	0.232	0.141	0.165	**0.678**	−0.045	−0.317
Participation of experts	**0.353**	0.265	0.273	**0.468**	0.024	0.186
Reliability of scientific knowledge	0.187	−0.136	−0.058	0.053	**0.848**	−0.071
Presence of prior successful cases within the administration	0.125	0.209	0.278	0.104	**0.652**	0.005
Experience with accidents and disasters	−0.050	0.015	0.019	−0.102	0.042	**0.789**
Eigenvalue	3.03	1.50	1.46	1.31	1.20	1.14
Percentage of cumulative distribution explained (%)	11.7	23.4	34.4	45.2	55.8	64.3

barrier of "lack of experience" shows a tendency to have a negative effect. These results are common sense and therefore consistent with our expectation.

Next, for disaster prevention departments, "disaster vulnerabilities" have the strongest effect and the perception of vulnerabilities increases the number of measures implemented. On the other hand, there are no apparent effects of the risk perception of external forces or situations to be avoided. The effects of drivers and barriers are strong for disaster prevention departments: it is clear that "securing budgeting" is also important as well as planning departments. One difference with planning departments is that the barriers of "external opposition" and "shortage of knowledge" increased the number of measures implemented. It surmises that the presence of cases, in which the number of measures implemented, increased as a result of proactive attempts due to an awareness of obstacles such as an external opposition or a shortage of knowledge.

Lastly, for environmental policy departments, similar to the analysis results of disaster prevention departments, the evaluation of vulnerabilities has a positive

Table 5.10 Results of principal component analyses of policy formulation barriers

Policy formulation barriers	f1	f2	f3	f4	f5	f6
Failure to secure subsidies from national government, prefecture, related foundations, etc.	**0.823**	−0.057	−0.014	0.117	0.031	−0.112
Lack of prior models by other local governments	**0.749**	0.163	−0.015	−0.018	−0.096	0.015
Lack of exchange of information with national government, prefecture, etc.	**0.732**	−0.096	0.162	−0.092	0.143	0.140
Lack of lively discussion	0.085	**0.887**	0.057	0.081	−0.086	0.017
Lowness of interest by businesses and industry organizations	−0.110	**0.779**	0.018	0.209	0.129	−0.170
Unreliability of scientific knowledge	0.037	**0.541**	−0.119	−0.011	**0.308**	0.018
Opposition of citizens' groups and nonprofits/NGOs	−0.124	−0.039	**0.904**	−0.046	−0.030	−0.117
Opposition of local residents	**0.328**	0.020	**0.835**	0.000	0.055	0.246
Lack of support by experts	−0.011	0.226	0.019	**0.739**	−0.096	−0.114
Shortage of experience and specialization within the administration	−0.114	0.045	−0.082	**0.705**	0.110	**0.389**
Difficulty of adjusting budgetary measures within the administration	**0.309**	−0.019	−0.049	**0.520**	0.143	**−0.426**
Lowness of interest among council and its membership	−0.113	0.208	−0.036	−0.117	**0.839**	−0.019
Lowness of interest by the head, and absence of policies	**0.354**	−0.004	0.121	**0.368**	**0.660**	−0.060
Lack of experience with accidents and disasters	0.078	−0.103	0.048	0.027	−0.028	**0.871**
Eigenvalue	2.49	2.22	1.50	1.25	1.14	1.11
Percentage of cumulative distribution explained (%)	15.5	28.6	39.9	50.9	60.5	69.3

effect. The variable with the highest value is the driver of "presence of other local governments." This tendency is distinctive to environmental policy departments, suggesting that the presence of other local governments to refer to is important in promoting policy. In many cases, policies are promoted as local governments follow each other's example. In these cases, it is likely that the presence of networks of local governments would be important as an agent to promote policy transfer. This point will be discussed in greater detail in a later chapter. Moreover, similar to the analysis results for planning departments, "past experience/desires of residents" has a positive effect and "lack of experience" had a negative one, indicating that experience is recognized to be important. In addition, the adjusted R^2 value for environmental departments is lower than those for other departments. This is a result of the fewer measures implemented by environmental policy departments (i.e., about one-half of that by other departments, as shown in Table 5.4).

Table 5.11 Correlation between drivers and barriers and the number of measures prepared and implemented

	Internal/external proactivity	Participation of other organizations	Presence of other local governments	Securing budgeting	Existing knowledge	Past experience/desires of residents	Shallowness of external relations	Lowness of external interest	External opposition	Shortage of knowledge	Lowness of internal interest	Lack of experience
Total	−0.028	−0.047	0.099	0.320**	−0.090	0.505**	0.271**	0.164	0.110	0.117	−0.089	0.052
Planning	0.203	0.078	0.120	0.634**	−0.093	0.724**	0.390	0.631**	0.020	0.043	−0.305	−0.136
Disaster prevention	0.115	0.081	−0.067	0.395*	−0.025	0.207	0.301	0.137	0.316	0.446**	−0.084	−0.086
Environmental policy	−0.153	0.047	0.155	0.176	−0.140	0.307*	0.086	−0.013	0.033	0.058	0.063	−0.180

$* p < 0.05, ** p < 0.01$

Table 5.12 Results of multiple regression analyses on the number of measures prepared and implemented using perception/evaluation factors and drivers/barriers

	Factor	Planning	Disaster prevention	Environmental policy
Perception	Disaster risks			
	Environmental risks			
	Planning risks			
	Planning vulnerabilities			
	Disaster vulnerabilities		0.408**	
	Environmental vulnerabilities			0.330*
	Disaster situations			
	Planning situations			
	Environmental situations			
Drivers	Internal/external proactivity			
	Participation of other organizations	−0.258**	0.243+	
	Presence of other local governments			0.457**
	Securing budgeting	0.520**	0.340*	
	Existing knowledge			
	Past experience/desires of residents	0.548**		0.371*
Barriers	Shallowness of external relations			
	Lowness of external interest	0.263*		
	External opposition		0.228+	
	Shortage of knowledge		0.319*	
	Lowness of internal interest			
	Lack of experience	0.235**		−0.350*
Adj-R^2 value		0.941	0.526	0.313
Model significance		$F(5, 10) = 48.7$**	$F(5, 26) = 7.9$**	$F(4, 35) = 5.4$**
Sample size		15	31	39

+ $p < 0.10$, * $p < 0.05$, ** $p < 0.01$

In the model shown in Fig. 5.4, a path extends from factors related to disaster prevention departments, such as disaster risks and crisis situations, to the number of measures prepared and implemented. Results of the multiple regression analysis in Table 5.12 are also consistent with the tendency that the effects of the factor of disaster vulnerabilities are strong for disaster prevention departments. At the same time, it is likely that for the other two departments, the effects of drivers and barriers are stronger and that the background of their lower compatibilities with the model become clearer. However, for drivers and barriers, the number of resilience measures prepared and implemented has not been fully explained. As a result, further verification of variables is needed. At this point, the results differ between civil indicators and administrative indicators.

5.4 Conclusion

With the administrative indicators, by analyzing the questionnaire data collected from departments of planning, disaster prevention, and environment in local governments across Japan, we examine the framework of the policy model in terms of the cognitive structure of the local officials. Major findings are as follows:

First, significant differences are observed among departments for most risks, with the exceptions of "drought, depletion of water sources (water resources)," "heat waves, fierce heat," "red tides," and "harmful insects." Planning departments showed high scores on long-term situations in local communities such as population decreases and aging. Disaster prevention departments showed high scores on not just natural external forces but also situations such as human-caused accidents and wars; and environmental policy departments showed high scores on situations related to typical types of pollution as well as long-term situations such as biodiversity and greenhouse gases.

Second, among all 28 indicators concerning vulnerabilities within local communities and government offices, significant differences among departments are observed on 13 indicators, such as "presence of steep terrain," "presence of rare or endangered species," and "high percentage of elderly population, depopulation," accounting for less than one-half of the number of indicators prepared. Unlike in the cases of risk perceptions discussed above, many of these indicators are common across local governments and departments, extending beyond the differences in departments' frameworks, which may result in the observed relatively fewer discrepancies among different departments.

Third, significant differences are observed among departments for most situations, with the exceptions of "long-term worsening of water resources," "worsening of living environment," and "long-term instability in energy supply." Planning departments again showed high scores on long-term situations in local communities such as "long-term decline in economic activity," "long-term decrease in the level of administrative services," "chronic worsening of public safety," and "degeneration of local culture/traditions"; disaster prevention departments showed high scores on severe circumstances in a disaster including "direct damage to human life"; and environmental policy departments showed the highest scores on "increase in global warming." These results strongly reflect the policy framework (the division of responsibilities) of each department.

Fourth, three overall tendencies can be identified for preparation and implementation of resilience measures: (i) the rates of transformative measures preparation and implementation are markedly low. Among these measures, the most frequently cited one was "development/support of next-generation energy infrastructure," followed by "centralization of urban functions (realizing a compact city)." (ii) although all departments reported implementing similar adaptative measures, the percentage of disaster prevention departments citing such measures was much higher than those of environmental policy departments and planning departments. (iii) Differences in frameworks were apparent among departments in terms of precautionary measures: 100% of disaster prevention departments reported the "formation of disaster agreements" and nearly 100% of environmental policy departments reported that they were "promotion of energy conservation."

Fifth, when examining the correlative relationships between the number of resilience measures prepared and implemented and some factors in each department, the following tendencies can be identified: (i) as for planning departments, the factors of the risk perception and evaluation of vulnerabilities and situations to be avoided are not significant, indicating that the effects of drivers and barriers in the policy process are stronger than those of these perception factors. The drivers of "past experience/desires of residents" and "securing budgeting" have particularly strong effects. (ii) As for disaster prevention departments, "disaster vulnerabilities" have the strongest effect and the perception of vulnerabilities increases the number of measures implemented. On the other hand, there are no apparent effects of the risk perception of external forces or situations to be avoided. The effects of drivers and barriers are also strong especially "securing budgeting" again. (iii) As for environmental policy departments, the variable with the highest value is the driver of "presence of other local governments." This tendency is distinctive to environmental policy departments, suggesting that the presence of other local governments to refer to is important in promoting policy. In many cases, policies are promoted as local governments follow each other's example. In these cases, it is likely that the presence of networks of local governments would be important as an agent to promote policy transfer. This point will be discussed in greater detail in a later chapter.

As a whole, the effects of drivers and barriers in policy process are stronger to the number of measures prepared and implemented. However, even for drivers and barriers, the number of resilience measures prepared and implemented has not been fully explained. As a result, further verification of variables is needed. At this point, the results differ between civil indicators and administrative indicators.

Acknowledgment We would like to express our sincere gratitude to the respondents who participated in the questionnaire survey and to Ms. Natsumi Ebitani (Graduate school of Agricultural and life sciences, University of Tokyo) and Ms. Izumi Hirata (Graduate school of Engineering, University of Tokyo) who assisted editing.

References

Adger W (2000) Social and ecological resilience: are they related? Prog Hum Geogr 24:347–364
Allen KM (2006) Community-based preparedness and climate adaptation: local capacity building in the Philippines. Disasters 30(1):81–101
Baba K (2005) Policy making processes on environmental and energy measures for sustainable city by Japanese local governments. J City Plann Inst Jpn 40(3):931–936. (in Japanese)

Baba K, Tagashira N, Jin Z (2012) A potentiality, drivers and barriers of diffusion of low carbon policy for business and industrial sector in local governments. J Environ Sci Jpn 25(2):73–86. (in Japanese)

Baba K, Matsuura M, Kudo T, Watanabe S, Kawakubo S, Chujo A, Tanaka H, Tanaka M (2017) Climate change adaptation strategies of local governments in Japan, Oxford research encyclopedia of climate science, 27 September, 2017, 1–27

Bergamini N, Blasiak R, Eyzaguirre P, Ichikawa K, Mijatovic D, Nakao F, Subramanian SM (2013) Indicators of resilience in socio-ecological production landscapes (SEPLs), UNU-IAS Policy Report. UNU-IAS. Retrieved August 5, 2015, from https://www.bioversityinternational.org/fileadmin/user_upload/online_library/publications/pdfs/Indicators_of_Resilience_in_Socio-ecological_Production_Landscapes_SEPLs_1676.pdf

Cutter SL, Barnes L, Berry M, Burton C, Evans E, Tate E, Webb J (2008) A place-based model for understanding community resilience to natural disasters. Glob Environ Chang 18:598–606

Godschalk D (2003) Urban hazard mitigation: creating resilient cities. Nat Hazard Rev 4:136–143

ICLEI-Local Government for Sustainability (2012) Preparing for tomorrow strategy 2012–2018, from http://archive.iclei.org/documents/LACS/Publicacoes/BH_Strategic_Plan_final_.pdf

Jha AK, Miner TW, Stanton-Geddes Z (eds) (2013) Building urban resilience: principles, tools, and practice. International Bank for Reconstruction and Development/The World Bank, Washington, DC

Longstaff PH, Armstrong NJ, Perrin K, Parker WM, Hidek M (2010) Building resilient communities: a preliminary framework for assessment. Homeland Secur Aff 1(3):1–23

Norris FH, Stevens SP, Pfefferbaum B, Wyche KF, Pfefferbaum RL (2008) Community resilience as a metaphor, theory, set of capacities, and strategy for disaster readiness. Am J Community Psychol 41:127–150

Sharma D, Singh R, Singh R (2013) Urban climate resilience: a review of the methodologies adopted under the ACCCRN initiative in Indian cities. Asian cities climate resilience working paper series 5. Retrieved October 14, 2015, from http://pubs.iied.org/pdfs/10650IIED.pdf

Tobin GA (1999) Sustainability and community resilience: the holy grail of hazards planning? Environ Hazard 1:13–25

UNISDR (2012) How to make cities more resilient a handbook for local governments leaders. UNISDR. Retrieved August 29, 2013, from https://www.unisdr.org/files/26462_handbookfinalonlineversion.pdf

van Aalst MK, Cannon T, Burton I (2008) Community level adaptation to climate change: the potential role of participatory community risk assessment. Glob Environ Chang 18:165–179

Part II
Case Studies

Chapter 6
Application of the Policy Model in Sendai: the Experiences of the Sendai Framework for Disaster Risk Reduction Action UNISDR/WCDRR Public Forum, 2015–2030

Kenshi Baba, Kosuke Shirai, Yu Nagata, Shun Kawakubo, Motoko Kosugi, Mitsuru Tanaka, and Ebru A. Gencer

Abstract This chapter reports on the raising of awareness among citizens who participated in the workshop on resilient cities held at the 3rd World Conference on Disaster Risk Reduction in Sendai in March 2015. We provided 18 citizens with four types of expert knowledge: a long-term climate forecast for Sendai; the roles of infrastructure as precautionary and adaptive measures; the roles of green infrastructure as adaptative and transformative measures; and the status report which consists of the assessment result of the resilience of Sendai City by urban indicators, civil indicators, and administrative indicators. Based on this expert knowledge, the participants discussed risk perception, vulnerability assessments, evaluation of situations to be avoided, and resilience measures in Sendai. The results demonstrate that

K. Baba (✉)
Tokyo City University, Yokohama, Kanagawa, Japan
e-mail: kbaba@tcu.ac.jp

K. Shirai (✉)
Mitsubishi Research Institute, Inc., Tokyo, Japan

Y. Nagata (✉)
Kumamoto City Office, Kumamoto, Japan

S. Kawakubo · M. Tanaka
Hosei University, Machida, Tokyo, Japan
e-mail: kawakubo@hosei.ac.jp; mtanaka@hosei.ac.jp

M. Kosugi
Shizuoka University, Shizuoka, Japan
e-mail: kosugi.motoko@shizuoka.ac.jp

E. A. Gencer
CUDRR+R Center for Urban Disaster Risk Reduction and Resilience 12 W, New York, NY, USA
e-mail: ebru.gencer@cudrr.org

M. Tanaka, K. Baba (eds.), *Resilient Policies in Asian Cities*,
https://doi.org/10.1007/978-981-13-8600-8_6

this workshop was able to change the participants' attitudes toward terms of self-support due to the fact that participants who had been personally affected by a disaster knew the importance of self-support and specific actions over public support.

Keywords Workshop · Expert knowledge · Local knowledge · Awareness raising

6.1 Introduction

"How safe is safe enough"(e.g., Lewis 2014)? This is the fundamental question of risk management—and yet disparities seem to remain between the risk information that has been quantitatively evaluated and presented by experts and the popular perception of risk. Whatever the nature of the problem, it will never be possible to eliminate risk completely, and when it comes to issues involving a great many uncertainties such as climate change, how to define acceptable risk and prioritize introduced policies in the political process are important. Particularly in scenarios when the public does not have the information necessary to make its own decisions about the risks it faces, the trust to risk managers or those who provide risk management information, such as experts or policy makers, becomes important (e.g., Earle and Cvetkovich 1995). Consequently, a dialogue needs to take place between experts and the public that uses the knowledge and values of both sides to close the gap between the results of risk analyses and the public perception of risk. In other words, it is hoped that integrating scientific knowledge from experts (expert knowledge), community-based knowledge gained from a comprehensive understanding of local realities, and other factors (local knowledge) will promote mutual understanding and social learning between actors (e.g., Irwin and Wynne 1996). We know it from various experiences of participatory tools such as the consensus conferences (e.g., Joss and Durant 1994) and scenario workshops (Bruyas and Riccio 2012).

In consideration of the above environment, this chapter will take the framework introduced in previous chapters and use the results of resilience evaluations gained via urban, civil, and administrative indicators to create status reports. This chapter will then examine what sort of results can be expected when using this approach as a means of strengthening and expanding a dialogue that, as stated above, will bridge the gap between risk analyses results and risk perception. Specifically, it will use the city of Sendai as a case study, analyzing the results of a workshop organized by the authors at the 3rd UN World Conference on Disaster Risk Reduction held in Sendai in March 2015, in which experts on the topic of "resilient cities" and the residents of Sendai City both participated. This chapter will report on the extent of awareness and education regarding Sendai's sources of risk and vulnerability, situations to be avoided, and resilience measures. This is a trial case study in Sendai with the status report we mentioned in Chap. 1.

6.2 Methodology

The workshop was held at Sendai City Hall on March 14 from 13:00 to 17:00 as a public forum at the 3rd UN World Conference on Disaster Risk Reduction. The workshop was entitled "Responding to the various urban risks studied at our workshop: Toward the creation of a resilience city model for Sendai." Sendai City was one of the cities targeted by the web questionnaire for citizens featured in Chaps. 3 and 4; respondents who expressed interest in the workshop were recruited, considering factors such as age, gender, place of residence, and disaster experience. However, as only a small percentage of a sample of approximately 440 Sendai citizens expressed interest in participating, additional participants were recruited from a panel conducted by a research agency; these additional participants had not taken the web survey. As a result, of the 18 participants, 70% were web survey respondents.

Furthermore, in order to use the external forces of climate change and natural disasters as a gateway to study resilience, the study showcased four subjects, detailed below, as its specialized fields of knowledge. In other words, for a total of 2 h, input was received from Sendai's district meteorological observatory on long-term climate observations for Miyagi and Sendai; from academics specializing in civil engineering and disaster prevention on the role of disaster prevention infrastructures in precautionary and adaptive measures; from academics specializing in agriculture and ecology on the role of ecosystem infrastructure in adaptive and transformative measures; and from the authors on the results of administrative indicators and civil indicators as they involve risk, vulnerabilities, situations to be avoided, and resilience measures. Furthermore, for the topics presented by the authors only, materials and documents were mailed out prior to the workshop. An outline of the workshop's agenda is presented in Fig. 6.1.

Afterward, participating citizens were divided into three groups and, with the help of a facilitator, discussed the following three topics for 30 min each: "What are the risks facing Sendai?"; "What are Sendai's vulnerabilities (weak points)?"; and "What can we do to make Sendai a more resilient city?" Post-It Notes recording individual participants' ideas and opinions were affixed to imitation vellum and shared; the facilitators collected the opinions in a more formal format. Finally, the citizen representative for each group presented a summary of the results (Figs. 6.2 and 6.3).

6.3 Results and Observations

6.3.1 An Overview of Sendai City Status Report

The assessment result of the resilience of Sendai City by urban indicators described in status report was as follows: (i) Sendai has higher defensive capacity and recovery capacity than big cities in Japan and has lower learning capacity. In particular,

Fig. 6.1 Workshop flyer distributed in Sendai City

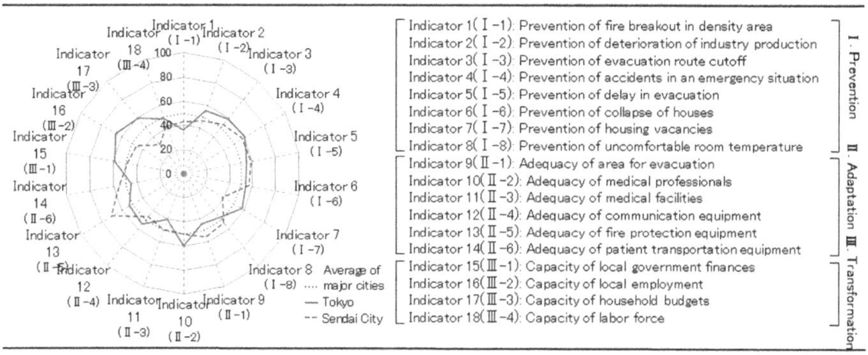

Fig. 6.2 Assessment result of Sendai City by urban indicators (1)

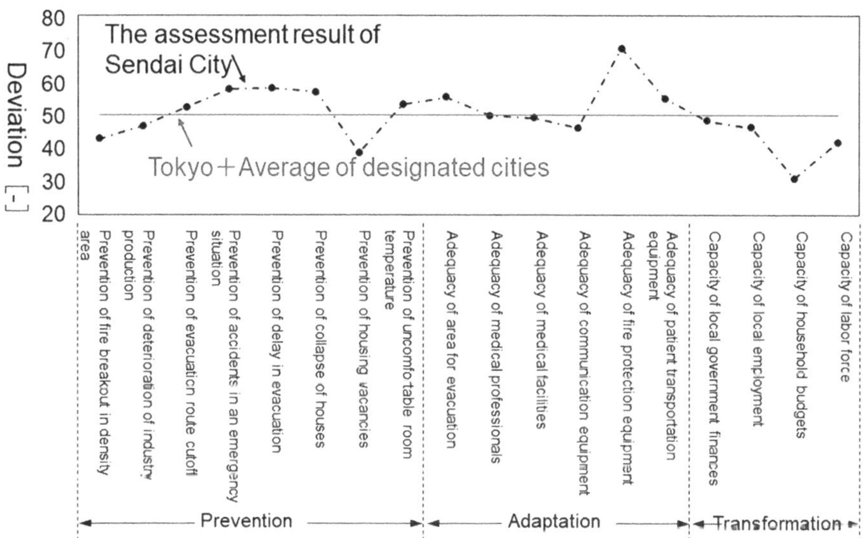

Fig. 6.3 Assessment result of Sendai City by urban indicators (2)

the fact that number of firetrucks and fireboats per capita is greater means that the initial response in a time of disaster is ready enough than the else big cities in Japan, whereas average saving rate which means that proportion of net increase of savings for disposable earnings is lower than the big cities in Japan which leads to an issue of coping with the cost of rebuilding at recovery period in each households.

As discussed in Chaps. 3, 4, and 5, the civil indicators and administrative indicators of Sendai City have the following characteristics.

First, Fig. 6.4 shows the average evaluations aligned on a 5-point scale (ranging from "not perceiving that it could be a crisis" to "perceiving that it could be a fairly serious crisis") for 41 indicators of risk awareness of natural forces and external

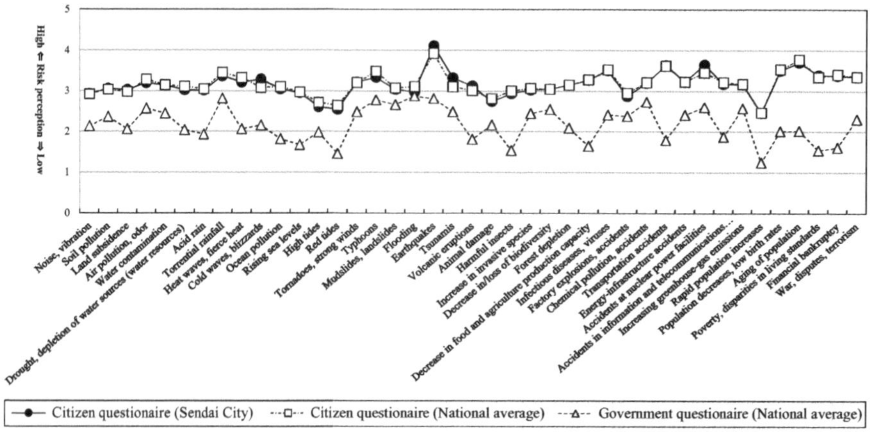

Fig. 6.4 Citizen's and local governments' risk perception caused by external forces (risk/stress)

social forces. This figure displays average values for the citizens of the nine regions of Japan versus the averages for the residents of Sendai City; it also represents an administrative indicator composed of national averages for local government survey data, collected separately. The figure shows nearly no differences between the risk awareness of the residents of Sendai and those of citizens as a whole. In Sendai, however, "earthquakes," "tsunamis," "accidents at nuclear power facilities," and "cold waves, blizzards" are rated more highly, while "heat waves, fierce heat" and "typhoons" are rated fairly low. When compared to the averages for local governments across Japan, represented in the figure as the administrative indicator, the citizens evaluate all phenomena as having higher risks than the local government does. The disparity was the smallest when it came to "flooding" and largest when it came to "traffic accidents," "poverty, disparities in living standards," "financial bankruptcy," and "aging of population."

Second, Fig. 6.5 shows the average response values aligned on a 5-point scale (ranging from "not vulnerable at all" to "very vulnerable") for 28 indicators used to assess vulnerability. Again, almost no discrepancies are evident between the vulnerability assessments of the residents of Sendai City and those of the citizens of the nine regions of Japan. However, in Sendai, a larger number of residents evaluate "presence of earthquake zones or fault lines" as a danger. Naturally, for every phenomenon, the public perceives a greater vulnerability than does the local government. The discrepancy is particularly wide for "high percentage of low-income earners, poor employment conditions" and "presence of active volcanoes."

Furthermore, when asked to what extent you should try to avoid the situation for 24 indicators, there was almost no difference between the evaluations of the residents of Sendai City and those of the citizens of the nine regions of Japan, and no remarkable trends were observed. Naturally, for all phenomena, more members of the public rated the given phenomenon as "to be avoided" than did members of the government. The gap was particularly large when it came to "cut-off of financial

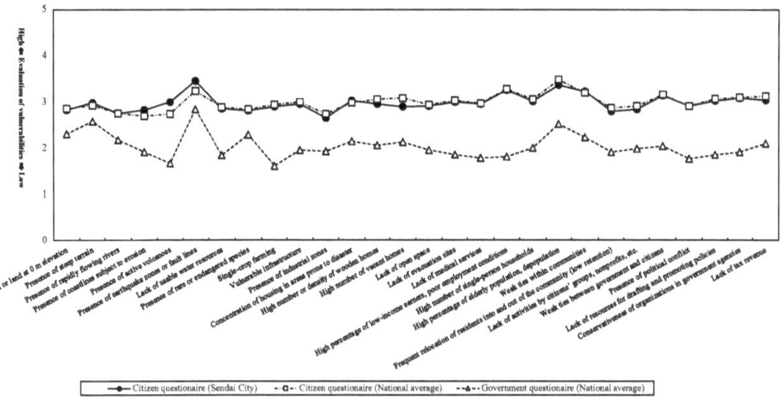

Fig. 6.5 Citizen's and local governments' assessments of vulnerabilities within local communities and households

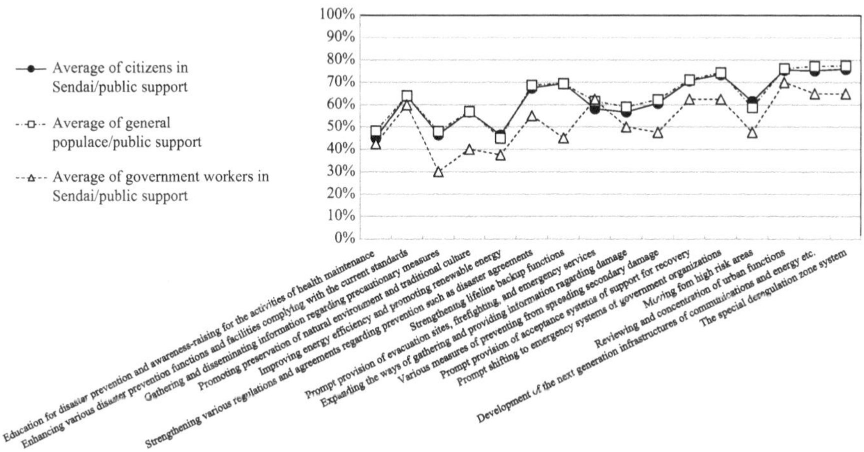

Fig. 6.6 Citizen's and officials' evaluation for resilience measures as public support

service functions," "chronic worsening of public safety," and "long-term decline in economic activity." However, there was almost no difference in assessments when it came to "direct damage to human life."

Third, Fig. 6.6 shows the percentage of people who evaluate public support (support and/or acceptance of government policies and activities) as appropriate for various resilience measures capable of providing support for risks, vulnerabilities, and situations to be avoided. Almost no difference is apparent between the evaluations of the residents of Sendai City and those of the citizens of nine regions of Japan, with a majority expecting public support for many resilience measures. Expectations are particularly high for "reviewing and concentration of urban functions" and "the special deregulation zone system." In addition, unlike the data in Figs. 6.4 and 6.5, the data in Fig. 6.6 cannot be directly compared with data collected into an administrative indicator; therefore, from the responses gathered into the civil indicators, the data from 400 respondents from the general public and 40 government workers living in Sendai City were taken and divided, with the latter data being used as a representative indicator for the administrative indicator. According to this data, an even greater number of government workers than everyday Sendai citizens expected public support in the cases of almost all resilient measures. Their support was particularly strong for "gathering and disseminating information regarding precautionary measures" and "strengthening lifeline backup functions." Almost no difference was apparent for "prompt provision of evacuation sites, firefighting, and emergency services."

Similarly, Fig. 6.7 displays the percentage of people who evaluate community support (proactive participation and cooperation via regional activities such as neighborhood councils and associations) as appropriate. For many resilient measures, those who expect community support total about 20–30%, with a greater number of Sendai government workers expecting it than the citizens of the nine regions of Japan. This trend is seen particularly strongly in regard to "gathering and disseminating information regarding precautionary measures" and "strengthening lifeline backup functions."

As expressed above, the impact of public perception of external risks, vulnerabilities, and situations to be avoided on evaluations of how to pursue resilience measure initiatives was examined in Sendai City's civil indicators via covariance structure analysis. The results are shown in Fig. 6.8 (reproduced of Fig. 3.8 in Chap.3). One remarkable feature of the residents of Sendai City is that the goodness of fit is stronger not for the path from disaster risks to environmental events, but for the path from social risks to environmental events. Also striking is the significant

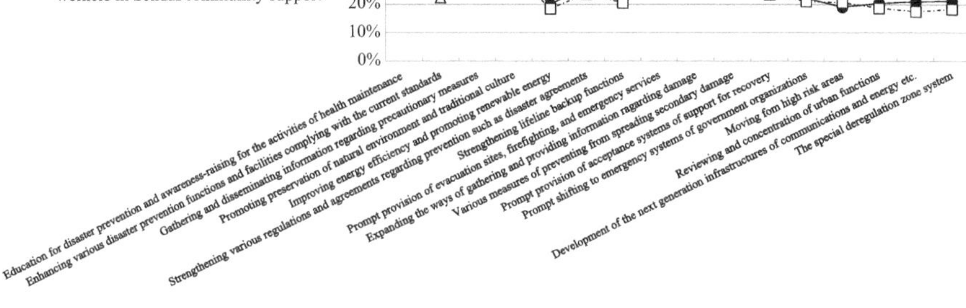

Fig. 6.7 Citizen's and officials' evaluation for resilience measures as community support

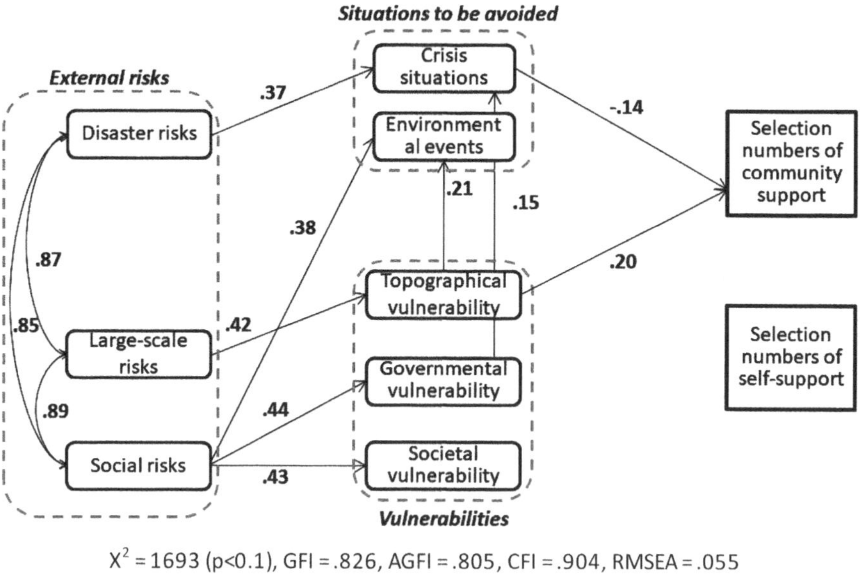

Fig. 6.8 Estimation of the cognitive structure with the civil indicators (Sendai City)

path from governmental vulnerability to crisis situations. In Sendai City, the positive impact from topographical vulnerability to community support is especially great. It is possible that the residents of Sendai grasp the importance of community support for vulnerabilities involving the presence of easily-eroded beaches and lowlands due to their experience of the Great East Japan Earthquake.

Furthermore, when it was examined whether there are any trends in the manner in which the eight types of disaster risks are understood by using "severity" and "unfamiliarity" as the map's axes, it was found that compared to other regions, Sendai City citizens clustered in the second quadrant, where disaster risk recognition is severe and familiar (Fig. 6.9, reproduced of Fig. 4.1 of Chap. 4). This suggests that the residents of Sendai trend relatively strongly toward being aware of many disaster risks as something severe yet familiar.

Indicators related to social capital such as commitment to the neighborhood and participation in local community activities are considered to represent resilience when it comes to reviving and rebuilding after disasters. On the other hand, awareness of vulnerabilities and presumptions of dangerous phenomena among neighborhoods and families are considered to signify resilience as it relates to taking prior precautions and countermeasures. From this perspective, the residents of Sendai rank somewhat low when it comes to the former factor, resilience in revival and restoration, but relatively high when it comes to resilience in prior planning.

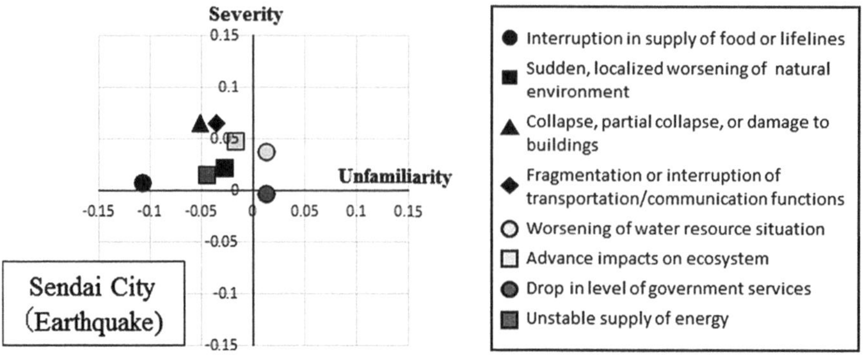

Fig. 6.9 Map of perceptions of disaster risk in each region of residence (Sendai City)

6.3.2 Workshop Results

As explained above, some of the distinctive characteristics of Sendai City were showcased by the authors in the workshop and included in the documentation mailed out in advance. As mentioned previously, expert knowledge was also provided on long-term climate projections for Miyagi and Sendai; the role of disaster prevention infrastructures in precautionary and adaptive measures; and the role of green infrastructures in adaptive and transformative measures. The workshop was then conducted. The results are presented below. The primary opinions are summarized in Tables 6.1 and 6.2.

6.3.2.1 Risk Perception

Regarding risk perception, the opinions and statements offered by Group 1 could be largely divided into the subjects of infrastructure (soft & hard), natural phenomena, social phenomena, climate change, and earthquakes and tsunamis. Specifically, in reference to Sendai City, when it came to infrastructure, the participants mentioned how disaster response is difficult in the suburbs due to the nature of a motorized society and how there are numerous hills and narrow roads but few alternate routes. Regarding natural phenomena, the participants mentioned the Mt. Zao volcano, the damage done to forests by landslides, and how service on the Tohoku Main Line railway had been frequently interrupted by strong winds. From a societal perspective, the striking aging of the population and sense of a disparity in volunteering between urban areas and the coastline were mentioned.

Broadly classified, the opinions and statements offered by Group 2 mentioned "nuclear power," "financial issues," "ground & foundation vulnerabilities," and

Table 6.1 List of opinions mentioned in the workshop

Group	Risks	Vulnerabilities
1	Infrastructure (hard & soft): Motorized society makes providing emergency support difficult in the suburbs Many hills, narrow roads; few alternate routes Natural phenomena: Eruptions of Mt. Zao Tsunami damage to forests Social phenomena: Severe aging of society; disparities among cities and coasts Tough to get around without a car (particularly for the elderly) Climate change Earthquakes & tsunamis	Infrastructure (hard & soft): Weak support during snowfall (Water services, city gas, waterways) Natural phenomena Social phenomena: Public transportation difficult to negotiate for elderly without a car Subway/bus fare is expensive Public transportation inconvenient in the suburbs (forced to rely on cars) Local human relationships weak for those who travel frequently for business Various local factors (city is geographically too long) Presence of nuclear power plants Climate change Earthquakes & tsunamis
2	Earthquakes: Disaster-mitigation forests washed away by tsunami Nuclear power plants: Presence of Onagawa Nuclear Power Plant Impact of radiation from Fukushima Nuclear Power Plant Financial issues Ground & foundation vulnerabilities: Ground vulnerabilities (created residential areas) Sudden torrential rains/ meteorological disasters Imbalances in the ecosystem: Pest & animal damage Other: Aging (two people dying in 2 years in apartment complex) Rotting of zelkova trees lining streets Volcanic activity of Mt. Zao	Composition of city/aging society: Broad urban area, low population density; difficult to create a compact city Weak sense of community in city center Infrastructure: High risk of flooding Weak public transportation due to motorized society Ground & foundation vulnerabilities: Ground weaknesses in filled/created land for housing developments presence of active faults Financial issues: Operation in financial deficit condition Lack of major companies Lack of employment Other: Eruptions of Mt. Zao

(continued)

Table 6.1 (continued)

Group	Risks	Vulnerabilities
3	Earthquakes Destruction of ground Destruction of infrastructure Natural disasters: Eruptions of Mt. Zao Warming Ecological changes due to warming Other: Presence of oil refineries & industrial complexes Radiation from resumption of operations at Onagawa Nuclear Power Plant Secondary disasters such as risk of shock from fallen/broken power lines	Personal level: Lack of preparation for evacuation Community level: (Aging) Government structures: Appropriateness of disaster maps and designated shelters (people have actually died in shelters; near a station is good for those not familiar with a given place) Infrastructure vulnerabilities: Deteriorating infrastructure (particularly city gas) Amount of auto traffic in cities; frequency of accidents Distribution of single-track lines Population flow Vulnerabilities in natural environment

Table 6.2 List of opinions mentioned in the workshop (cntd)

	What is required to create a resilient city?			
Group	Self-support	Community support	Public support	Other
1	Disaster preparedness: Being proactive in obtaining information Preparing food & water Securing furniture Securing a method of contacting one's children Ecologically-conscious driving Electricity conservation Obtaining earthquake insurance, etc. Elevating disaster awareness: Improving disaster awareness	Creating & strengthening a sense of community Creating events & groups and planning communication to strengthen weakening human relationships	Transportation infrastructure: Expansion of transportation Dissemination of official information: Expansion of communication infrastructure Provision of information Exchange of information between officials and public: Development of disaster prevention leaders	

(continued)

Table 6.2 (continued)

	What is required to create a resilient city?			
Group	Self-support	Community support	Public support	Other
2	Improved awareness of information: Use city newsletters, etc., to increase awareness of information Improved awareness of self-assistance Improved awareness of self-assistance	Local communication: Creating structures by which to gather & share information discovered by citizens during periods when the government is not mobilized or disseminating information Creating & strengthening a sense of community: Creating & strengthening a local sense of community	Gathering & providing accurate information during emergencies: Gathering, providing, & disseminating accurate information, such as sites with risk and earthquake & tsunami damage during emergencies Requests for assistance to other areas: Requesting assistance from other areas Creating structures that can actively provide information to the government from the public: Structures that actively transmit information from the public to the government Developing awareness of self-assistance Disseminating information to increase awareness of self-assistance Effective use of the legislative system Effective use of the legislative system and special wards	Diffusion of renewable energy Application of green infrastructure Public introduction of the latest technology Appropriate distribution of donation funds Maintenance of places to think (maintenance of places to consider how best to create a resilient city) Attracting research facilities

(continued)

Table 6.2 (continued)

Group	What is required to create a resilient city?			
	Self-support	Community support	Public support	Other
3	Voluntary danger avoidance: Avoiding dangerous areas as a general rule Preparation for disaster (disaster-prevention inspections etc.): Confirmation of evacuation routes Cooperating in practicing evacuation procedures Participation in disaster prevention training (participating in disaster-prevention inspections, evacuation procedure practice, etc.): Teaching children about disaster prevention As a specific idea, creating a memorial day to remember the experience of March 11 & conduct disaster-prevention inspections etc.	Local communication: Communicating with neighbors; gathering & storing information about those who live within a certain radius Creating & strengthening a sense of community: Creating disaster-prevention leaders on neighborhood councils and workplace disaster-prevention organizations Publishing local newsletters Creating evacuation manuals: Creating manuals for evacuation procedures for each area Mobilizing the young (university students, etc.): Attempting to mobilize university students during disasters	Expansion of green infrastructure: Preparing for emergencies; increasing the number of trees lining the streets Creating an emergency support manual: Creating a support manual for emergencies Thorough provision & maintenance of evacuation areas: Provision & thorough maintenance of evacuation areas	Development of disaster prevention awareness: Conducting tours of disaster-struck areas; promoting disaster awareness

"earthquakes" as having a great impact on daily life. "Sudden torrential rains/meteorological disasters" and "imbalances in the ecosystem" were frequently mentioned as having a low impact on daily life. Specifically, the presence of the Onagawa Nuclear Power Plant and the impact of radiation from the Fukushima plant were mentioned in connection with "nuclear power," and the expense of post-earthquake restoration was mentioned in connection with "financial issues." For "the makeup of the city/graying society," the deaths of two people in 2 years in managed apartment complexes and the presence of vulnerabilities in the Earth's foundation were mentioned; for climate change, the effect of damage from pests and wildlife was mentioned; and for "other," activity at Mt. Zao was mentioned.

The opinions and statements offered by Group 3 can be broadly categorized into the presence of "earthquakes"; the resultant destruction of buildings and collapse of mountains; the "destruction of ground" from flooding, etc.; "destruction of infrastructure" such as the breaching of embankments, the collapse of highways, and the severing of lifelines; "natural disasters" such as tornadoes/waterspouts, climate change, and torrential rains; "ground collapse," such as the collapse of developments; and, furthermore, fires and radiation from nuclear plants. Specifically, the eruption of Mt. Zao and ecosystem changes due to warming were cited in connection with "natural disasters," and the presence of oil refineries and industrial complexes, as well as secondary disasters such as the danger of shocks from severed power lines, were cited in connection with "other."

From the above, all groups mentioned "earthquakes," "volcanic activity at Mt. Zao," "climate change," "societal aging," and "nuclear power."

6.3.2.2 Assessing Vulnerability

Regarding evaluations of vulnerabilities, the opinions and statements offered by Group 1 could be largely divided into not only "earthquakes & tsunamis" and "climate change," but also "social phenomena," such as the reduction in human interaction and the increase in apartment complexes, and "infrastructure" issues related to public transportation and lifelines. Specifically, besides the presence of nuclear power plants, Group 2 respondents cited as "social phenomena" the fact that human connections in the country are weak from the perspective of transferees from Tokyo, how the city of Sendai spans a geographically wide area, and the city's lackluster capacity for coping with snow; for "infrastructure," they cited the difficulty of negotiating public transportation for the elderly who have no cars, and the high subway and bus fares (Fig. 6.10).

Fig. 6.10 Scenes from the workshop

When the opinions and statements offered by Group 2 were classified, many respondents ventured the opinion that the social phenomena of city structure and the aging society might have an impact on transportation infrastructure, and that the slump in local industry might have an impact on the generation currently raising children. For potential issues, ground vulnerabilities, the presence of active faults, and agriculture adapting to climate change were also mentioned. Specifically, for "city composition/aging society," the difficulty of creating a compact city due to large urban areas and low population density, the high risk of flooding after snowfall, and public transportation being weak in a motorized society were mentioned on the hard infrastructure side, and a weak inner-city community, the lack of major business, the lack of employment, and the city running on a deficit budget were mentioned on the soft infrastructure side. For "potential issues," vulnerabilities in filled land and residential developments and eruptions at Mt. Zao were cited.

The opinions and statements offered by Group 3 could be largely divided into "the individual level," "the community level," "government systems," "vulnerabilities in the infrastructure," "population flow," and "vulnerabilities in the natural environment." Specifically, for "the individual level," the low level of disaster-prevention awareness and the lack of preparation for evacuation were cited; for "vulnerabilities in the infrastructure," deteriorated infrastructures (city gas), the number of vehicles and traffic accidents in the city, and the weaknesses in single-line distribution were cited. For "government systems," the lack of a solid disaster-prevention manual and the inadequacy of disaster maps and designated shelters (in which refugees have actually died) were cited.

Of the above, all groups cited "a weak sense of local community," "societal aging," "transportation inconveniences (the motorized society)," "the vast size of the urban area," and "infrastructure (city gas, etc.)."

6.3.2.3 Expectations for Self-support, Community Support, and Public Support

As for expectations of self-support, community support, and public support when it comes to resilience measures, Group 1's comments can be broadly classified into "preparation," "communication," "infrastructure (transportation)," "awareness," and "government & citizen information." When classified into self-support, community support, and public support, the following findings were made. For self-support, Group 1 cited the preparation of food and water, securing one's furniture, guaranteeing a means of contact with one's children, environmentally conscious driving, electricity conservation, obtaining earthquake insurance, etc., and being proactive in obtaining information; "disaster preparation" to improve awareness of disaster prevention was also mentioned. For community support, "creating & strengthening a sense of community" through planning communication and creating events and groups to strengthen weakened human relationships was mentioned. For public support, the expansion of transportation and communication infrastructures, training disaster prevention leaders, and the provision of information were mentioned.

In Group 2, for self-support, using tools such as city newsletters and raising information awareness, as well as elevating awareness of self-support, were mentioned; for community support, "local communication" and "creating and strengthening a sense of community" via creating a system for collecting and sharing information through citizen sites at times when the government isn't mobilized to provide information, and creating and strengthening a sense of local community for that purpose were mentioned. For public support, "gathering and providing accurate information in emergencies" through obtaining, gathering, and disseminating accurate information in emergencies, such as areas where risk and tsunami damage are present; requesting support from other areas; promoting awareness of self-support; and effective use of the legislative system and special wards, etc., were mentioned. Also cited were the diffusion of renewable energy; the application of green (ecologically-friendly) infrastructure; the introduction of the latest technology among the public; and the appropriate distribution of donation funds.

In Group 3, for self-support, "voluntary crisis avoidance" (as a general rule, staying away from dangerous places), "disaster preparation" (confirming evacuation routes, cooperating with disaster drills), and "participating in disaster prevention training" (teaching children about disaster prevention and heightening disaster awareness through specific ideas such as holding a day of remembrance for March 11 and conducting disaster inspections) were mentioned. For community support, "local communication" through communicating with neighbors and gathering information about those living within a certain radius; "creating and strengthening a sense of community" through starting workplace disaster prevention organizations, inaugurating neighborhood disaster prevention leaders, and publishing local newsletters; standardizing evacuation procedures for each neighborhood; and making sure university students can mobilize in a disaster were mentioned. For government assistance, "expansion of green infrastructure" (providing emergency support and increasing the number of trees on the roadside), creating a response manual for various disasters, and provision and thorough maintenance of shelters were mentioned. Other measures such as holding tours of disaster-struck areas to develop disaster awareness were also mentioned.

For self-support, multiple groups mentioned "elevating self-support awareness and awareness of information" and " disaster preparation" as self-support measures. As for community support, while the specific methods suggested differed, multiple groups mentioned "creating & strengthening a sense of local community" and "local communication." For public support, most participants mentioned "providing accurate information," "using the legislative system," as well as "developing infrastructures" such as transportation, information, and green infrastructures.

6.3.3 Awareness Among Participating Citizens as Viewed Through a Post-workshop Questionnaire

A survey was conducted among participants after the end of the workshop. Based on their responses when asked to specify "areas where I want experts and the government to be more involved" and "areas where I really want to be more involved" in making Sendai a resilient city, respondents (N = 16) were classified (Table 6.3) according to their level of awareness. Their responses were then further sorted into the categories "increasing self-support awareness and acting," "creating and strengthening a sense of local community," "participating in workshops and events," "disseminating information and raising information awareness," "disaster preparedness," and "other." When comments and statements regarding expectations for self-support, community support, and public support offered when discussing the workshop were compared, their content was not found to differ by category, but it demonstrated a desire not only to initiate independent efforts to disseminate information such as "disseminating disaster information via blogs" and participate in disaster-prevention events such as "gaining the knowledge and leadership to be able to lead others during disasters" but also to adopt a proactive approach through "being one of those who provide information and first aid to others."

Actions that respondents wanted experts and the government to take to make Sendai a more resilient city were classified into "legislative system & policies," "developing infrastructure; contributing to the introduction of green infrastructure," "gathering, providing, and sharing information between the public and government," "pressure on government by experts," "increasing awareness of self-support," and "other."

The mention of "experts work[ing] together across disciplines" indicates the possibility that having experts from various fields simply enumerate the claims of their respective disciplines might not lead to public understanding or a public consensus. Therefore, it might be necessary for experts to become more aware of, and pay more attention to, the means by which they provide expert knowledge, such as visualizations like issue mapping and influence diagrams. Also, the references to self-support might suggest the possibility of a change in attitude. Regarding this, the authors present as a topic how the general public has even greater expectations for community and public support than it does for self-support. It is suggested, then, that the findings of this study do not necessarily agree with those of public opinion polls from the Japanese Cabinet Office (2015) stating that self-support is viewed by the public as important. The importance of self-support, primarily among those who have previously been the victims of disasters, is also definitely identified.

The community disaster-prevention plans of local governments in Japan have been formed by simply bundling together measures related to disaster prevention and natural disaster reduction extracted from the existing administrative plans of relevant departments (Nagamatsu 2008). Properly speaking, the priorities of measures should be established in the planning process through integrating expert knowledge and local knowledge by stakeholder participation such as this workshop.

Table 6.3 Results of post-workshop questionnaire

Areas where I want experts and the government to be more involved	Areas where I really want to be more involved
Legislative system & policies: Taihaku Ward seems to be home to more children than the other wards; I want policies conducive to an increasing population Natural disasters are unavoidable, but the soft infrastructure (the ability to help each other locally) needs to be strengthened to cope, and the Personal Information Protection Act is an obstacle to that	Increase awareness of self-assistance and acting: Mental readiness Constant awareness Somehow increase awareness of self-assistance Get a little more self-awareness, then act Expand my own mental readiness as well as that of my family, neighborhood, and other areas & wards Rethink what I can do and refocus on putting it into action Be constantly aware of small abnormal events and take action before they become big problems
Developing infrastructure; considering the introduction of green infrastructure: Research on and consideration of introducing a green infrastructure Development of transportation (development of irrigation channels; snow removal; flooding)	Creating & strengthening a sense of local community: Building a community, particularly through individual initiatives Hold events in residential areas
Gathering, providing, and sharing information between the public and government: Complete provision of information from the government I want one-on-one exchanges of ideas between private citizens I want information to be provided on private initiatives in which I can take part over a wide area Creating a structure where citizens come together on a regular basis to bring up problems and consider ways to solve them	Participating in workshops & events: Help create forest wind buffers I want to participate in workshops Gain the knowledge and leadership to be able to lead others during emergencies
Pressure on government by experts: I want them to demand more from the government I want experts to work together across disciplines and pressure the government Proposals from experts to Sendai's mayor, city council, and the Miyagi prefectural governor	Increasing information awareness; transmit information Share earthquake & disaster experience on blogs Read city newsletters
Increasing awareness of self-assistance: Repeatedly practice self-assistance Strongly instill in children an awareness of "70% self-assistance" Communicate the idea of "70% self-assistance" to adults as well	Disaster preparedness Purchase evacuation supplies Check designated evacuation areas, too Prepare for earthquakes
Other: Coordinate with other towns & prefectures during emergencies Guarantee funding	Other Talk with my family today These projects should be conducted at elementary & middle schools, too

In the future, it is highly possible that the changing climate might lead to magnified or more frequent natural disasters. For example, if the public is not involved in determining the exact issues and challenges involved with the risk of climate change right at the ground level—if the public does not participate, and if experts and the public do not share knowledge of the scientific facts and awareness—mutual understanding will never develop, and agreement on the priorities of resilience measures will not be reached. The public workshop attempted by this study proactively provides an opportunity to examine adaptive risk management from the bottom up and seeks change in the manner in which local government plans are drafted and devised (such as the introduction of risk management methods).

6.4 Conclusion

This chapter analyzed the results of a workshop on resilient cities held with experts and citizens of Sendai City at the 3rd UN World Conference on Disaster Risk Reduction, and collected and summarized knowledge and experiences on the extent of awareness of and learning of Sendai City's risks, vulnerabilities, situations to be avoided, and resilience measures with the status report measured by urban, civil, administrative indicators.

As a preparation, a web questionnaire survey was conducted in nine regions across Japan, including Sendai, which we have described in Sect. 12.3. The participants were recruited based on their age, gender, and disaster experience, as well as an interest in the workshop. Eighteen participants were selected. We provided the participants with four types of expert knowledge: a long-term climate projection for Sendai, the roles of infrastructure as precautionary measures, the roles of green infrastructure as adaptative and transformative measures, and the results of assessment with the three indicators. Based on this expert knowledge, the participants were divided into three groups and discussed risks, vulnerability, situations to be avoided, and resilience measures in Sendai. The main findings are as follows:

First, most participants mentioned "elevating self-support awareness and awareness of information" and "disaster preparation" as self-support measures. As for community support, while the specific suggested methods differed, most participants mentioned "creating & strengthening a sense of local community" and "local communication." For public support, most participants mentioned "providing accurate information," "using the legislative system," as well as "developing infrastructures" such as transportation, information, and green infrastructures.

Second, the participants tended to demonstrate greater expectations for public support and community support than self-support initially, but during the workshop, participants with disaster experience of the Great East Japan Earthquake indicated the importance of self-support, resulting in the attitude change of most participants. This workshop encouraged the change in participants' attitudes from public support to self-support against long-term risks. It means that this type of workshop can provide opportunities to examine the adaptive risk management in a bottom-up approach and to seek a change in the manner of planning and risk management, involving both the citizens and the local governments.

Acknowledgment We would like to thank Professor Toshimitsu Komatsu (Kyushu University), Ms. Yukiko Ikeda (Sendai district meteorological observatory), and Professor Toru Nakashizuka (Tohoku University) who presented their expert knowledge at the workshop, and also Professor Noriko Sugiyama (Nagoya University), Dr. Naoki Masuhara (Research Institute of Humanity and Nature), Dr. Haruna Ozawa (The Coalition of Local Governments for Environmental Initiative, Japan), Mr. Kosuke Shirai (Mitsubishi Research Institute, Inc.) and Mr. Yu Nagata (Kumamoto City Office) who conducted the workshop collaboratively with us, and the citizen in Sendai who attended the workshop. We would also like to thank Ms. Natsumi Ebitani (Graduate school of Agricultural and life sciences, University of Tokyo) and Ms. Izumi Hirata (Graduate school of Engineering, University of Tokyo) who assisted editing.

Appendix; Sendai Framework for Disaster Risk Reduction DRR) 2015–2030 and the New Ten Essentials of Making Cities Resilient

Introduction

Representatives from 187 UN member States adopted the first major agreement of the Post-2015 development agenda in Sendai, Japan, on 18th March 2015. The Sendai Framework for DRR, which builds on the Hyogo Framework of Action aims to achieve "the substantial reduction of disaster risk and losses" over the next 15 years. This outcome will be achieved by the goal to: "[p]revent new and reduce existing disaster risk through the implementation of integrated and inclusive measures that prevent and reduce hazard exposure and vulnerability to disaster, and increase preparedness for response and recovery, and thus strengthening resilience" (UN 2015).

To support the assessment of global progress, seven global targets have been agreed to make substantial changes in the:

1. Reduction of global disaster mortality by 2030
2. Reduction of the number of affected people by 2030
3. Reduction of direct disaster economic loss in relation to global GDP by 2030
4. Reduction of disaster damage to critical infrastructure and disruption of basic services by 2030
5. Increase in the number of countries with national and local DRR strategies by 2020
6. Increase in the availability and access to multi-hazard early warning systems and disaster risk information and assessments by 2030
7. Enhancement in international cooperation to developing countries through adequate and sustainable support by 2030

This Post-2015 DRR Framework is further supported by guiding principles and four priorities for action.

The Four Priorities of Action

Priority 1. Understanding Disaster Risk
Disaster risk management should be based on an understanding of disaster risk in all its dimensions of vulnerability, capacity, exposure of persons and assets, hazard characteristics, and the environment. Such knowledge can be used for risk assessment, prevention, mitigation, preparedness, and response.

Priority 2. Strengthening Disaster Risk Governance to Manage Disaster Risk
Disaster risk governance at the national, regional, and global levels is very important for prevention, mitigation, preparedness, response, recovery, and rehabilitation. It fosters collaboration and partnership.

Priority 3. Investing in Disaster Risk Reduction for Resilience
Public and private investment in disaster risk prevention and reduction through structural and nonstructural measures are essential to enhance the economic, social, health, and cultural resilience of persons, communities, countries, and their assets, as well as the environment.

Priority 4. Enhancing Disaster Preparedness for Effective Response and to "Build Back Better" in Recovery, Rehabilitation, and Reconstruction
The growth of disaster risk means there is a need to strengthen disaster preparedness for response, take action in anticipation of events, and ensure capacities are in place for effective response and recovery at all levels. The recovery, rehabilitation, and reconstruction phase is a critical opportunity to build back better, including through integrating disaster risk reduction into development measures.

The Role of Local Governments and the New Ten Essentials for Making Cities Resilient

During the World Conference in Disaster Risk Reduction in Sendai, the role of local governments as the primary responsible authority in disasters was strongly emphasized. Local and subnational governments argued for a greater need for international collaboration, and reiterated their commitment to collaborate with national governments and with the global community to advance the implementation of the Sendai Framework for DRR 2015–2030. Further, in a declaration, local and subnational requested the continuation of "Making Cities Resilient: My City is Getting Ready" Campaign with added support and direct assistance at local level.

To support the then 2500+ cities that are part of the Making Cities Resilient program, a new set of "Ten Essentials" and subsequent indicators were developed by a team of experts and organizations. The primary objective of these essentials is to be operational, adaptive, and applicable to all, encouraging cities towards their implementation.

The New "Ten Essentials" build upon the previous Essentials, and are interlinked to the Sendai Framework for DRR with priorities for action, representing a transition to a stage of implementation.

It is expected that the Sendai Framework and the New Ten Essentials will pave the way for implementing new policies on urban resilience and support governments in implementing strategies that include targets and are time bound. The New Ten Essentials for Making Cities Resilient[1] are:

1. Organize for Disaster Resilience
2. Identify, understand, and use current and future risk scenarios
3. Strengthen financial capacity for resilience
4. Pursue resilient urban development and design
5. Safeguard natural buffers to enhance the protective functions offered by natural ecosystems
6. Strengthen institutional capacity for resilience
7. Understand and strengthen societal capacity for resilience
8. Increase infrastructure resilience
9. Ensure effective disaster response
10. Expedite recovery and build back better

References

Bruyas A-M, Riccio M (2012) Science centres and science events: a science communication handbook. Springer, Dordrecht

Earle TC, Cvetkovich G (1995) Social trust: toward a cosmopolitan society. Praeger Press, Westport

Irwin A, Wynne B (eds) (1996) Misunderstanding science? The public reconstruction of science and technology. Cambridge University Press, Cambridge

Joss S, Durant D (1994) Consensus conferences: a review of the Danish, Dutch and UK approaches to this special from of technology assessment of the options for a proposed Swiss consensus conference, The science museum library

Lewis EE (2014) How safe is safe enough?: technological risks, real and perceived. Carrel Books, New York

Nagamatsu S. (2008). An approach to theory of disaster risk reduction, Koubundou publisher. in Japanese

The Cabinet Office of Japanese Government, (2013) A public-opinion poll on disaster prevention (survey in December, 2013), http://survey.gov-online.go.jp/h25/h25-bousai/index.html in Japanese

United Nations, (2015) Sendai framework for disaster risk reduction 2015–2030, Geneva: UNSIDR, https://www.preventionweb.net/files/43291_sendaiframeworkfordrren.pdf

[1] Available at http://www.unisdr.org/campaign/resilientcities/home

Chapter 7
Multi-criteria Evaluation of Local Energy Resilience

Tsuneo Takeuchi

Abstract Based on the data from the Great East Japan Earthquake, I have assessed the prevention (e.g., recovery from blackouts), adaptation (e.g., restoration of thermal power stations), and conversion (conversion into distributed energy systems) with the multi-criteria of the resilience value (e.g., the cost of avoiding blackouts), CO_2 emissions reduction, and capital expenditure. As a result, based on these three criteria, conversion into distributed energy systems is the most desirable approach for local energy resilience.

Keywords Environmental resilience · Energy resilience · Decentralized energy system · Climate change · Regional revitalization

7.1 Environmental Resilience

The Fifth Assessment Report (2014) by Working Group II of the Intergovernmental Panel on Climate Change (IPCC) identified the following eight risks of extreme weather events associated with climate change: (1) sea level rise, storm surges, and coastal flooding; (2) flooding of inland and urban areas; (3) systemic risk; (4) periods of extreme heat; (5) food insecurity; (6) insufficient access to water in semi-arid regions; (7) loss of marine and coastal ecosystems; and (8) loss of terrestrial and inland ecosystems.

Similarly, the Japanese government's Fundamental Plan for National Resilience (Cabinet Decision of May 2014) identified several environment-related risks to be addressed in case of large-scale natural disasters, such as damage of energy supply, drinking water supply, and sewage networks; failure of sewage and waste treatment facilities; large-scale leakage and spreading of hazardous materials; and destruction of farmland and forests. The Fundamental Plan for National Resilience should also address the climate change risks defined by the IPCC, except for heat waves.

T. Takeuchi (✉)
Graduate School of Environmental Studies, Nagoya University, Nagoya, Aichi, Japan
e-mail: tsuneo_takeuchi@nagoya-u.jp

M. Tanaka, K. Baba (eds.), *Resilient Policies in Asian Cities*,
https://doi.org/10.1007/978-981-13-8600-8_7

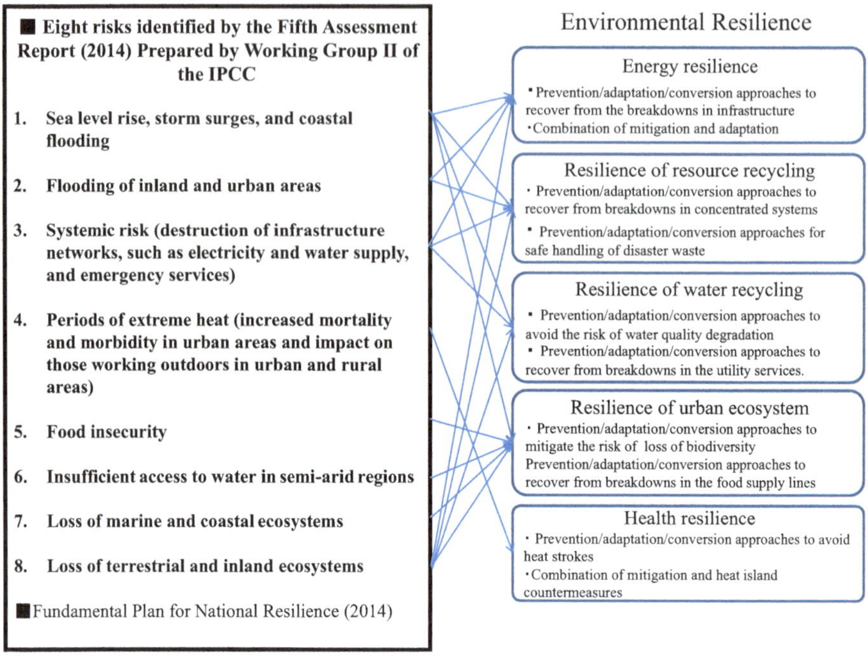

Fig. 7.1 The eight risks identified by the IPCC and the corresponding environmental resilience. (Source: Created by the author according to the IPCC)

Therefore, five areas [i.e., regional energy systems, resource recycling systems, water recycling systems, urban ecosystems, and health (the risks of heat stroke)] require an environmental policy to mitigate the risks of anthropogenic climate change and large-scale natural disasters.

In this chapter, responses of these sectors were referred to as "environmental resilience" and measures taken to support these responses were referred to as "environmental resilience measures" (Fig. 7.1).

7.2 Energy Resilience: Conversion into Distributed Energy Systems

The third risk identified by the IPCC Fifth Report is "Systemic risks due to extreme weather events leading to breakdown of infrastructure networks and critical services, such as electricity, water supply, and health and emergency services." Up to now, this field has received little attention in terms of adaptation. Extreme weather events can cause destruction of the electricity infrastructure, such as power generation and transmission facilities, associated with an impact on the supply systems, such as electricity and water networks, as well as other services, such as hospitals, ambulances, and fire-fighting. Society is increasingly becoming dependent on electricity, especially with the recent advances in the Information and Communication Technology (ICT) and the recent focus on the Internet of Things (IoT). The electricity demand is quality oriented, i.e., electricity resilience is prioritized to quantity. Therefore, energy resilience is an important aspect of environmental resilience.

Energy systems, particularly regional and urban energy systems, should simultaneously be self-adaptive and able to relief the adverse effects of climate change. Therefore, it is becoming necessary that these systems integrate into mutually beneficial relationships.

Counter measures taken to mitigate the effects of human-caused climate change and large-scale natural disaster on the energy sector, i.e., energy resilience measures, can be broadly categorized into "prevention" (e.g., reinforcing networks), "adaptation" (e.g., rapid recovery of damaged facilities), and "conversion" (i.e., conversion from traditional to distributed energy systems).

First, these energy resilience measures were evaluated using the following indices: "resilience value" (e.g., the cost of avoiding blackouts), "CO_2 reduction volume," and "resilience capital investment amount." These indices were then used to quantitatively evaluate and assess each measure according to its category (i.e., prevention, adaptation, and conversion) taking the household and commercial sectors in Nagoya City as case studies.

7.2.1 *Resilience Value*

To assess the resilience value (e.g., the cost of avoiding blackouts), we established quantitative scenarios based on data, such as the time period before thermal power, gas, water supplies are restored; period needed for the restoration of thermal power stations; and length of planned blackouts, from the Great East Japan Earthquake (Table 7.1). Three scenarios were proposed: prevention (decreasing the period required to restore utilities by installing suitable equipment, such as redundant power lines), adaptation (reduce the planned blackout periods by reducing the period required for the restoration of thermal power stations), and conversion (reducing the period required to recover from blackouts resulting from the destruction of transmission lines via converting to distributed energy systems and replacing city gas by waste heat from hot water streams). To calculate the respective expenses (value), the quantitative scenarios were multiplied by the respective planned blackout cost and the savings in the city gas charges (Takeuchi 2016).

Two prevention quantitative scenarios were proposed: (1) reducing the period required for a 90% recovery from blackouts by 1 day (from 5 to 4 days) through installing redundant power lines, and (2) reducing the period required to recover from city gas cuts by 5 days (from 25 to 20 days) by implementing several measures, e.g., replacing the existing pipes by polyethylene (PE) gas pipes.

Table 7.1 Recovery periods in the Great East Japan Earthquake used for evaluating energy resilience measures

Number of days required for the restoration of thermal power stations (February 2017, Japan Society of Civil Engineers), dividing the total number of days required for the restoration of each power station by the number of power stations	127 days
Number of days required for 90% recovery from a blackout (Tohoku Electric Power service area)	5 days
Number of days before recovering from water supply cuts (Sendai City Waterworks Bureau)	16 days
Number of days before recovering from city gas cuts (Sendai City Gas Bureau) (light damage to pipelines and main pipeline network)	25 days
Number of planned blackouts per household (Tokyo Electric Power Co.)	2.5 times
Duration of the planned blackout (Tokyo Electric Power Co.)	4 h
Cost of planned blackout response (large enterprises) ["Survey Report on Blackout Cost" (January 2014, Electric Power System Council of Japan)]	¥1919/kWh
Cost of planned blackout response (SMEs) ["Survey Report on Blackout Cost" (January 2014, Electric Power System Council of Japan)]	¥7497/kWh
Cost of planned blackout response (individuals) ["Survey Report on Blackout Cost" (January 2014, Electric Power System Council of Japan)]	¥1431/kWh
Tohoku Electric Power Co. 2011–2013 capital investment for disaster recovery	¥212.0 billion

Source: Created by the author

The resilience values for each prevention scenario (e.g., the cost of avoiding blackouts) were calculated as follows: for scenario (1), the following formula was used: total household electricity consumption (kWh) × cost of individual blackout response (¥1431/kWh) + total commercial electricity consumption (kWh) × cost of small and medium-sized enterprise (SME) blackout response (¥7497/kWh)) × (24 h/8760 h), and for scenario (2), the following formula was used: total amount of energy required for household and commercial hot water supply (MJ) × 5 days × ¥4.3/MJ. The calculated values were (1) = ¥18,353 million, (2) = ¥1641 million, with a total value of ¥19,968 million for both scenarios.

One adaptation scenario was proposed: (3) reducing the period required for the restoration of thermal power stations from 127 to 90 days, i.e., reducing the duration of planned blackouts from 10 h (2.5 × 4 h) to 7 h. The resilience value (cost of avoiding blackouts) was calculated as follows: total amount of household electricity consumption (kWh) × cost of individual planned blackout response (¥1431/kWh) + total amount of commercial electric power consumption (kWh) × cost of SME blackout response (¥7497/kWh)) × (3 h/8760 h). The value was calculated to be (3) = ¥22,943 million (Table 7.2).

Three conversion quantitative scenarios were proposed: (4) reducing the period required for the recovery from a blackout resulting from destruction of power transmission lines by 2 days (from 5 to 3 days) by covering 40% of the total household and commercial electricity consumption with retail sales from distributed electricity (e.g., cogeneration, private power generation, waste incineration generation, and solar power); (5) reducing the duration of planned blackouts from 10 h (2.5

Table 7.2 Resilience value, CO_2 emissions, and capital investment associated with prevention, adaptation, and conversion measures in the household and commercial sectors in Nagoya City (test calculation)

	Prevention	Adaptation	Conversion
Quantitative scenario	(1) Reducing the period required for a 90% recovery from blackouts by 1 day (reduction from 5 to 4 days)	(3) Decreasing the period required for the restoration of thermal power stations from 127 to 90 days (reducing planned blackouts from 10 h (2.5 × 4 h) to 7 h)	(4) Reducing the period required to recover from a blackout resulting from destruction of power transmission lines by 2 days (from 5 to 3 days) by covering 40% of the total electricity consumption of household and commercial sectors with retail sales from distributed electricity (cogeneration, home generation, refuse generation, solar power)
	(2) Reducing the period required to recover from city gas cuts by 5 days (from 25 to 20 days) by implementing several measures, such as installing PE pipes		(5) Shortening the time of planned blackouts from 10 h (2.5 times × 4 h) to 5 h by covering 40% of total electricity consumption of household and commercial sectors with retail sales from distributed electricity
			(6) Obtaining the total amount of thermal energy required for hot water needs of the household and commercial sectors from waste energy of factories, waste incinerators, and cogeneration (excluding a 14-day recovery period from water supply cuts)

(continued)

Table 7.2 (continued)

		Prevention	Adaptation	Conversion
Resilience value calculation formulas		(1) (total household electricity consumption (kWh) × cost of individual blackout response (\1431/kWh) + total commercial electricity consumption (kWh) × cost of SME blackout response (\7497/kWh)) × (24 h /8760 h)	(3) (total household electricity consumption (kWh) × cost of individual rolling blackout response (\1431/kWh) + total commercial electric power consumption (kWh) × cost of SME blackout response (\7497/kWh)) × (3 h /8760 h)	(4) (total household electricity consumption (kWh) × individual blackout response cost (\1431/kWh) + total commercial electricity consumption (kWh) × cost of SME blackout response (\7497/kWh)) × (24 h /8760 h)
		(2) Total energy required for household and commercial hot water supply (MJ) × 5 days × \4.3/MJ	–	(5) (total household electricity consumption (kWh) × cost of individual blackout response (\1431/kWh) + total commercial electricity consumption (kWh) × cost of SME blackout response (\7497/kWh)) × (5 h /8760 h)
				(6) total energy required for household and commercial hot water supply (MJ) × ((number of days required to recover from city gas cuts – number of days required to recover from water supply cuts)/365 days) × \4.3/MJ
Resilience value (\100 million)	(1) (3) (4)	18,355	22,943	36,709
	(5)			38,239
	(2) (6)	1614		2905
	Total	19,968	22,943	77,853

CO_2 emissions reduction (kiloton)	(1)	–	–	1020
	(3)			
	(4)			
	(5)	–	–	
	(2)	–	–	1370
	(6)			
	Total	–	–	2390
Capital expenditure for prevention, adaptation, and conversion		Capital investment for electric power system recovery: \48.4 billion to \56.4 billion (applying the capital investment calculated per kWh and per population to the recovery of Tohoku Electric Power to Nagoya City)		Capital investment for 200,000 kW cogeneration (excluding heat conduit) approx. \50.0 billion

Source: Created by the author

times × 4 h) to 5 h by the same action (i.e., converting to distributed power systems); and (6) obtaining the total thermal energy required for household and commercial hot water supply from the waste heat obtained from factories, waste incinerators, and cogeneration (excluding a 14-day period for the restoration of water supply).

The resilience values for conversion scenarios (4)–(6) were calculated as follows: (total household electricity consumption (kWh) × cost of individual blackout response (¥1431/kWh) + total commercial electricity consumption (kWh) × cost of SME blackout response (¥7497/kWh)) × (24 h /8760 h)), (total household electricity consumption (kWh) × cost of individual blackout response (¥1431/kWh) + total commercial electricity consumption (kWh) × cost of SME blackout response (¥7497/kWh)) × (5 h/8760 h)), and (total energy required for household and commercial hot water supply (MJ) × ((number of days required for recovering from city gas cuts – number of days required for water supply restoration)/365 days) × ¥4.3/MJ), respectively. The calculated values were: ¥36,709 million, ¥38,239 million, and ¥2905 million, respectively, with a total value of ¥77,853 million.

This analysis shows that the resilience values of the conversion scenarios are much larger (Fig. 7.2).

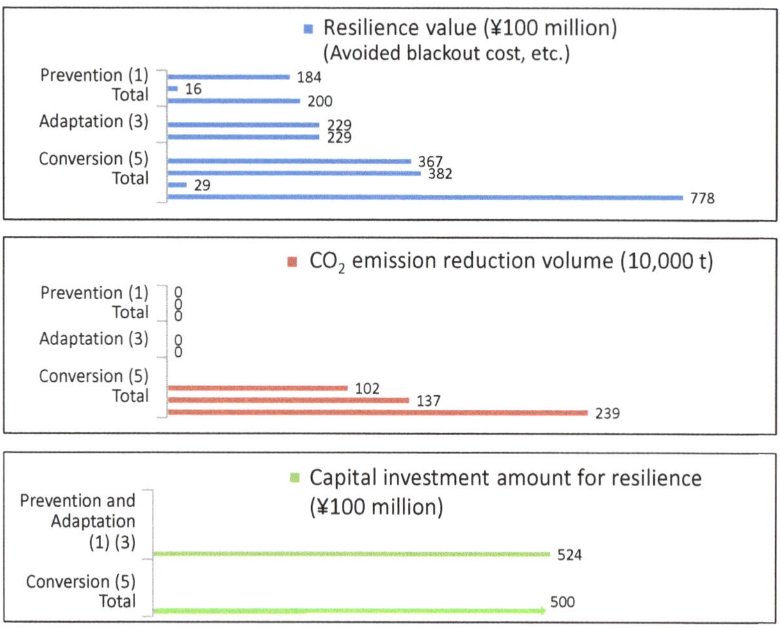

Fig. 7.2 Resilience values, CO_2 emissions, and capital investment associated with resilience measures. (Source: Created by the author)

7.2.2 Reduction of CO_2 Emissions

CO_2 emissions can only be reduced under the conversion scenarios. The total reduction in the CO_2 emissions under scenarios (4)–(6) is 2,390,000 tons.

7.2.3 Capital Investment for Resilience Measures

According to the "Initiatives for Streamlining Management" proposed by Tohoku Electric Power Co., Inc. (February 2013), the disaster recovery capital investment between 2011 and 2014 was ¥212.0 billion [was consumed in: main thermal power station (approximately 3,400,000 kW), 46 pylons, 105 transmission lines, 75 substations, and approximately 36,000 power poles]. Based on this information, we calculated the capital investment per electricity consumption (kWh) and per population for prevention and adaptation measures in the service area of Tohoku Electric Power Co., Inc., and applied the results to household and commercial electricity consumption and population in Nagoya City. This calculation yielded ¥48.4 billion based on electricity consumption (kWh) and ¥56.4 billion based on population.

The conversion capital investment (used for cogeneration equipment) was calculated to be around ¥50.0 billion with a rated output of around 200,000 kW (excluding heat conduit).

The conversion capital investment is almost equivalent to the total capital investment for prevention and adaptation measures.

Based on these results, the resilience value of conversion was found to be larger than the sum of those of prevention and adaptation, and the CO_2 emissions due to conversion was found to be reduced by around 15% for the total area of Nagoya City. The capital investment for conversion was found to be similar to the sum of those for prevention and adaptation. Conversion to distributed energy system was found to be higher in all indices than prevention (e.g., reinforcing networks) and adaptation (e.g., restoration of damaged facilities).

Conversion to a distributed energy system is clearly an effective strategy from the resilience and environmental, such as CO_2 reduction, perspectives.

In the past, the energy policy in Japan was not supportive of the conversion into distributed energy systems. However, there has been a progressive change in this regard, particularly since the Great East Japan Earthquake that resulted on the disaster of Daiichi Nuclear Power Station (operated by Tokyo Electric Power Company Holding, Inc.) (March 2011). With the introduction of an electricity purchasing system for renewable energy in July 2012, there has been a noticeable progress in the conversion of electric power system towards distribution systems, such as full liberalization of the retail electricity market starting in April 2016, and unbundling of the power transmission from the power distribution is planned to be achieved by 2020. The Noda cabinet's environmental and energy strategy (September 2012), which was formulated through a national discussion, stated that by 2030, 45% of the electricity sources would be distributed energy resources (30% from renewable energy and 15% from cogeneration). The target of Germany is to have 60% of its total electricity sources from distribution energy sources by 2020 (35% from renewable energy and 25% from cogeneration).

Further, in July 2015, the "Long-term Energy Supply and Demand Outlook" planned the 2030 electricity source composition to be around 27% from LNG, around 26% from coal, around 22–24% from renewable energy, around 20–22% from nuclear power, and around 3% from oil. This composition is based on fuel; therefore, the ratio of cogeneration is not mentioned. However, it is about 11%. Thus, in 2030, the contribution of distribution energy resources would be about 33–35% (about 22–24% from renewable energy and 11% from cogeneration). It must be noted that this energy policy not only runs counter to the global trend but also against common sense.

Furthermore, under the Basic Energy Plan, based on the Basic Act on Energy Policy, the four basic perspectives of energy policy are safety, stable supply, economic efficiency, and environmental suitability. These aspects have been firmly retained in the new "Long-term Energy Demand and Supply Outlook." However, energy resilience cannot be inferred from these basic perspectives.

Therefore, the results can be summed as follows: compared to prevention (e.g., reinforcing networks) and adaptation (e.g., rapid recovery of damaged facilities) measures, the most effective approach to energy resilience was found to be conversion to distributed energy systems. Moreover, this approach combines several measures to adapt to climate change and alleviate its effects into a single strategy. The measure proposed to enhance this effect would be discussed later, but first it was necessary to examine the impact of human-caused climate change on energy systems and the countermeasures to mitigate this impact (adaption measures).

7.3 Adaptation Measures to Achieve Energy Resilience in Germany

The German Federal Government laid out its first nationwide adaptation strategies in the 2005 Climate Protection Program. In addition, the Federal Environment Agency established the Competence Centre on Climate Change Impacts and Adaptation Center (KomPass) in 2006. In 2008, the Federal Government decided on the "German Strategy for Adaptation to Climate Change," and in 2011, an "Adaptation Action Plan" was adopted under the agreement of the Federal Government and State Government. The adaptation strategy and KomPass are being achieved through adaptation measures in 14 fields [health, construction, water/shoreline/marine ecosystems, soil, biodiversity, agriculture, forestry, fisheries, energy (conversion, transmission, and distribution), finance, transport, industry, tourism, and district/construction planning]. Germany's adaptation measures in the energy field are discussed in the following section.

It was decided within the adaptation framework that the infrastructures for energy conversion and distribution must be protected from extreme weather events such as floods, torrential rains, storms, hail, and lightning, while considering the impact of climate change on energy demand. Therefore, the framework clarifies the policy and the technological, legal, and management responses.

The first point in the framework focused on energy conversion, such as electricity generation, where it pointed out that the conventional energy supply system, which is based on concentrated, large-scale power stations, is vulnerable to damage from climate change, while converting the energy supply to renewables, in particular wind and solar power, distributes the energy supply and consequently increases the system resilience against climate change phenomena in these regions.

The main risk posed by climate change to conventional energy systems is related to cooling water supply for thermal power stations, such as coal, oil, natural gas, and nuclear power stations. A greater volume of cooling water is required when it is obtained from warmer rivers, because the higher temperature decreases its cooling effect. The standards set out by the current Wasser Gesetz (Water Act) already preclude the use of warmed up river water as cooling water. The practice is restricted by the Water Act, because it further increases the temperature of the water returned to the river, which causes problems in the river ecosystem. In extreme cases, power stations are forced to halt their operations. To avoid these problems, an innovative cooling concept combining both water and air in the cooling systems should be adopted for steam power stations. In addition, since cogeneration utilizes higher fuel temperatures and offers greater power generation efficiency, it should be used to benefit from the waste heat and reduce the halts caused by cooling water shortage. Converting into renewable energy generation can also be adopted as a countermeasure to the cooling water problem in the summer months, when the problem is more pressing, because it can reduce the peak load demand in fossil fuel power stations.

In addition, power stations and their related facilities are at risk of being inundated by flooding caused by climate change. Even with ensuring high levels of safety and implementing flood countermeasures in power stations, such as locating them on high grounds, damage and disconnection can occur, e.g., the accident in the Kraftwerk Gera Süd (Power plant Gera South) power station in 2013. Although today's concentrated power stations have drainage systems, their buildings are still exposed to floods; therefore, it is necessary to appropriately choose the sites of new power stations. Floods can also influence household heating equipment, and hence causing damage to boilers and hot water pipes.

Moreover, the logistics of conventional power stations had to be considered. For example, the ships that are used to transport fuel to conventional power stations may be hindered by flooding or low water levels, and in these cases, alternative means must be arranged. Therefore, it is necessary to adopt flexible logistics plan and study the possibility of stockpiling fuel at local sites.

It is very important that the impact of climate change on renewable energy is also considered. For example, an increase in temperature will reduce the efficiency of the photovoltaic modules, while hail and storms will cause damage. In addition, wind power generation needs to be halted in strong storms and it stops generating electricity during stable periods of high pressure, water ingress into a photovoltaic electricity system can cause short circuits and damage the modules, flooding can impair the production of biomass, and solar power generation is impaired when covered by snow. Storage of renewable energy (wind and solar powers) will have a significant role in the future as a countermeasure to these situations. Even today, the impact of storms, snow, and hail is significant. For example, a number of concepts and technologies are already being developed, such as a wind control device for wind power generation, to tackle this problem. Moreover, remote maintenance systems of wind power stations are being efficiently used. Solar power stations situated in areas at risk of flooding need to be equipped with switching systems to prevent short circuits. Hail damage to solar power modules can be prevented by using thick protective glass.

The second point in the framework was related to the stability of the electricity distribution network. In selecting the sites for the network facilities, it is essential to consider extreme weather events that could damage them. Floods increase the risk of damaging transformers and other network equipment. Cable routes can become flooded, damaging pylons and pylon foundations. Furthermore, flooding can also have an impact on the gas distribution network and remote energy supply networks; e.g., reformers and gas compressors. Terminal cables damages and transmission losses from overhead lines that can occur during the hottest summer periods can cause transmission faults. The load on the distribution network from the weight of snow or abnormal temperatures requires technical prevention measures. Monitoring the temperature and the overhead transmission network are necessary to reduce the risk on high transmission capacity equipment. Heater wires and pulse electro-thermal deicing (PETD) are effective for preventing freezing.

The third point considered within the framework is the impact on energy demand. As a result of higher temperatures, the cooling demand of the households, factories, and materials transport and storage will increase. This point in particular needs to be considered in building construction and urban planning. For example, building insulation and shade creation clearly reduce the cooling demand. Another measure is to increase the energy efficiency of technological equipment.

This above-mentioned information summarizes the various initiatives proposed by KomPass. However, the situation in Japan differs in some points; for example, obtaining cooling water for steam power stations from rivers, and the future growth in cooling demand. Germany, a leader in global warming alleviation measures, has conducted a multifaceted examination of adaptation measures in the field of energy, and is taking additional steps to further engage with the issue.

The Japanese government decided on the first "National Plan for Adaptation to the Impacts of Climate Change" in December 2015. The plan mainly focuses on adaptation measures for the agriculture, forestry, and fisheries sectors; water environment and resources; and natural ecosystems. The plan mentioned energy systems only once, stating that "Regarding energy demand and supply, due to the limited number of examples, it cannot be inferred that there is consensus regarding damage and impacts on the energy infrastructure due to an increase in the frequency and intensity of extreme events or long-term sea-level rise."

7.4 Regional Revitalization Effect of Regional Electricity Retail Business

We calculated the regional revitalization effect resulting from the introduction of regional electric power retail business, which is one of the distributed energy systems, on a model city (Fig. 7.3).

This model city has a population of 200,000 people and renewable electricity capacity of 360,000 kW based on solar and waste power generation. This investment was fully covered by the city. The upper section of Fig. 7.3 shows that electric power company A receives electricity fees of ¥11.5 billion from the city and purchases renewable electric power through a fixed price purchase scheme of ¥11.3 billion, which is paid to renewable energy investors. The balance of income and payments inside the region is –¥0.2 billion.

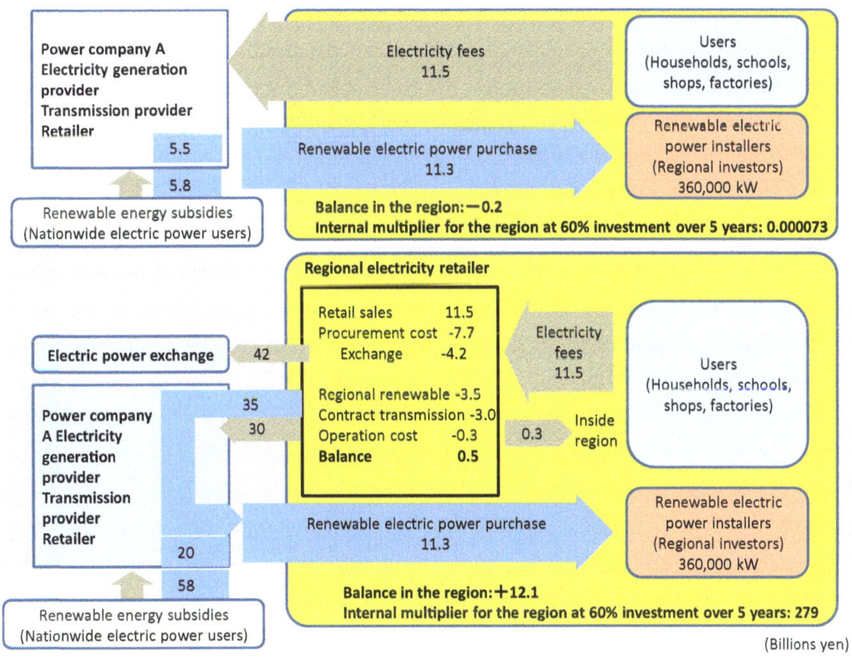

Fig. 7.3 Regional revitalization effect resulting from introducing regional electricity retail business (test calculation using a model city). (Source: Created by the author)

The lower section of Fig. 7.3 shows that a regional electric power retailer was established, and renewable electric power is purchased inside the region. The deficit is purchased from electric power wholesalers, and all the power is sold to retail users inside the region. An electricity fee of ¥11.5 billion is paid to the retailer. The retailer pays ¥4.2 billion as an exchange and ¥ 3.0 billion as a contract transmission fee to electric power company A. The renewable energy investors pay ¥11.3 billion in total, including the cost of renewable energy purchase (¥3.5 billion) from inside the region via electric company A. In this case, the balance of income and payments within the region is + ¥12.1 billion.

If 60% of this amount is spent and invested inside the region, and 60% of this 60% continues to be spent and invested locally over the ensuing years, the cumulative amount after 5 years would be ¥27.9 billion. The regional internal multiplier in this case would be 2.31.

7.5 Launch of the "Global Covenant of Mayors for Climate and Energy – Japan"

The "Global Covenant of Mayors for Climate and Energy – Japan" is scheduled to launch in 2018. Under the covenant, leaders of local governments will pledge to act on the three issues targeting the year 2030; namely (1) local production for local consumption of energy (promotion of renewable electric power, thermal cogeneration, and regional electric power retail business); (2) setting CO_2 reduction targets that are higher than the national targets for 2030; and (3) adapting to climate change.

Local governments that sign the pledge will acquire a new brand as a part of an international network of local governments that will support the recommendations of the Paris Accord at the regional level. The Paris Accord is an international framework to supersede the Kyoto Protocol starting in 2020.

The local governments that join the covenant will also become leaders in the regional revitalization movement through actions, such as local production for local consumption of energy.

In addition, the Mayor's Covenant is an important tool for increasing the energy resilience at the regional level.

7.6 Conclusion

"Environmental resilience" is a policy concept that deals with the ability to respond to risks such as climate change and large-scale natural disasters in an integrated manner. It was defined by evaluation of the risks in multiple areas such as regional energy systems, resource recycling systems, water recycling systems, urban ecosystems, and health (the risks of heatstroke).

In "energy resilience," its measures are categorized into "prevention," "adaptation," and "conversion." The multi-criteria evaluation for energy resilience measures was used by evaluation indices, such as the resilience value, CO_2 reduction amount, and the amount of resiliency equipment investment. Then, energy resilience was quantitatively evaluated, and measures were compared for prevention, adaptation, and transformation, targeting Nagoya city as a case study. As a result, it turned out that transition to a decentralized energy system is an effective measure not only from the viewpoint of the environment – such as CO_2 reduction and the reinforcement of energy sources – but also from the viewpoint of resilience.

Furthermore, the author also conducted a test calculation for model cities regarding the regional revitalization effects of regional electricity retail businesses in a decentralized energy system. In the case where a regional electric power retailer purchases renewable electric power by territory and sells to retail users in the region, most of the electricity fees paid out of the region remain, and internal multipliers increase significantly. This case shows that establishing a power retail business in a region and purchasing and selling the renewable energy in that territory contributes to regional revitalization.

Acknowledgment This research was supported by the Environment Research and Technology Development fund (1-1304) of the Ministry of the Environment, Japan.

References

Survey Report on Blackout Cost (2014) Electric Power System Council of Japan
Takeuchi T (2016) Expectations and suggestions to the community by the electricity market liberalization: from effect analysis to the local community and cases in Germany. Env Inf Sci 45(1):25–31
The Fifth Assessment Report (2014) Intergovernmental Panel on Climate Change (IPCC)

Chapter 8
Enhancing Capacities for Building Climate and Disaster-Resilient Cities in Asia: Case Study of Cebu, Philippines and Nonthaburi, Thailand

Premakumara Jagath Dickella Gamaralalage, Toshizo Maeda, and Simon Gilby

Abstract Enhancing capacity for building resilient cities is a growing concern among policy makers and international communities to minimize the impacts of climate change and natural disasters. The UN Conference on Sustainable Development 2012 and the Sendai Framework for Disaster Risk Reduction (2015–2030) called for urgent need in building resilient cities to face climate change and associated disasters. For that, city governments need to be aware of current and future potential risks and prepare in order to enhance the resilience of the urban systems and communities. Drawing on the experience of Cebu and Nonthaburi, this chapter aims to examine how developing cities can effectively plan and take specific measures to enhance resilient communities. First, it gives an overview of the recent evolution of theoretical and conceptual issues related to resilience in development practice. Based on these analyses, a conceptual framework that involved four key elements (context, disturbance, assets, and capacity to deal with disturbance, and reaction to disturbance) was developed to understand key factors influencing vulnerability and enhancing resilience of communities. Relevant information was gathered by literature review, key informant interviews, and focus group discussions with relevant stakeholders. The study findings suggested that resilience is a process rather than a static state, and as such, building resilience cities to deal with disturbance requires improving three distinct but interrelated capacities (absorptive, adaptive, and transformative), which are mutually reinforcing and exist at multiple levels. In planning and carrying out measures for achieving resilience, emphasis was given to the actions on disaster risk reduction (DRR), social protection, natural resource management, and management of public goods and services. For the effective implementation, city government need to enhance the potential for science-based policy making, effective leadership, efficient financing, multi-sector partnership, land use planning, and citizen participation.

P. J. Dickella Gamaralalage (✉) · T. Maeda (✉) · S. Gilby
Institute for Global Environmental Strategies (IGES), Kitakyushu, Japan
e-mail: premakumara@iges.or.jp; maeda@iges.or.jp; gilby@iges.or.jp

M. Tanaka, K. Baba (eds.), *Resilient Policies in Asian Cities*,
https://doi.org/10.1007/978-981-13-8600-8_8

Keywords Climate change · Resilient cities · Adaptation plan · Community participation · Enabling policies

8.1 Introduction

The rapid urban growth and climate change poses huge risks to quality of life, economic, and social stability of the cities, especially in developing countries where one out of seven people is living in informal settlements and urban slums (Mitlin and Satterthwaite 2013). Evidence suggests that climate change exerts added stress on cities bringing higher temperature, increase of sea level, intense rainstorms, droughts, inland and coastal flooding, landslides, and heat waves. It may also pose particular threats to urban infrastructure and basic services, such as water supply, energy provision, sanitation, transport, and ecosystem services (IPCC 2007, 2012; UCCRN 2011).

Furthermore, cities face a rising costs of natural disasters that have been increasing in frequency and intensity in over the last 10 years. According to UNISDR (2015), more than 700,000 people have died as a result of natural disasters and more than 1.4 million people have been injured. Around 23 million people have lost their homes. Some 1.5 billion people have been adversely affected by disasters, which have caused material losses amounting to USD 1.3 trillion. Between 2008 and 2012, 144 million people were displaced by disasters (UNISDR 2015). While mega disasters such as the Indian Ocean Tsunami (2004), the Great East Japan Earthquake and Tsunami in Japan (2011), larger-scale floods in Bangkok, Thailand (2011), the tropical storm in the Philippines (2013), and the earthquake in Nepal (2015) capture global attention, the cumulative impacts of every-day and small-scale disasters at the local level affect more people and are major impediments to long-term development of cities (World Bank and GFDRR 2013).

These realities have led to significant international commitment to enhance capacity for building resilient cities. The eighth UN Secretary-General Ban Ki-moon stated that "I call for the need of world leaders to address climate change and reduce the increasing risk of disasters – and world leaders must include Mayors, townships and community leaders" (Quote by Ban Ki-moon, the eighth UN Secretary-General at the Incheon Conference "Building an Alliance of Local Governments for Disaster Risk Reduction," Aug 2009). The Sendai Framework for Disaster Risk Reduction (2015–2030), a successor of the Hyogo Framework for Action (2005–2015), called for urgent actions on achieving disaster risk reduction and resilient cities (UNISDR 2015). The UN Conference on Sustainable Development – the Future We Want (2012) – also called actions for disaster risk reduction and building of resilience cities to disasters in achieving sustainable development and poverty reduction. The Report of the High-Level Panel of Eminent Persons on the Post-2015 Development Agenda emphasizes the importance of establishing human settlements adopting and implementing policies and plans towards resilience and adaptation to climate change and natural disasters (UN 2013).

As a responsible institution which is the closest to the citizens and play the first role in responding to crises and emergencies and in attending to the needs of their constituencies, local governments need to get ready, reduce the risks, and become more resilient to disasters. Further, local governments deliver essential services to their citizens (health, education, transport, water, etc.), which need to be made resilient to disasters. Evidence also suggests that investing in resilience programming that reduces exposure to risk is significantly more cost effective for local governments than post-disaster responses (WMO 2008).

Internationally, there are some ongoing efforts that focus on building capacity of cities and local governments to understanding risks and building resilient cities in achieving sustainable development. For example, the Making Cities Resilient Campaign of the UNISDR (United Nations Strategy for Disaster Reduction) that was launched in 2010 to advocate widespread commitment by local governments to build resilience to disasters and increased support by national governments to cities for the purpose of strengthening local capacities. The Asia Pacific Adaptation Network (APAN) established in 2009 has been functioning as a platform for sharing effective approaches for adaptation and facilitating access between development and investment plans and funding mechanisms. ICLEI-Local Government for Sustainability has been organizing Annual Global Forum on Urban Resilience & Adaptation since 2010 which convenes about 500 participants globally including leaders from local governments, representatives of international organizations, research institutions, and nongovernmental organizations. The 100 Resilient Cities – pioneered by the Rockefeller Foundation (100RC) – is another global initiative dedicated to helping cities around the world become more resilient to the physical, social, and economic challenges that are a growing part of the twenty-first century.

Reflecting these trends, drafting city development plans by integrating disaster risk reduction and climate change adaptation is getting importance in order to reduce current threats and emerging impact from climate change (Pelling 2003; Prabhakar et al. 2009). This increased interest not only presents opportunities for those promoting in disaster response and recovery, but also more importantly for long-term disaster risk exposure reduction and nonstructural hazard mitigation strategies. Despite this growing recognition, developing cities in Asia are not yet receiving adequate attention for introducing effective measures for planning resilient cities. Drawing on the experience of two Asian cities, including Cebu City, Philippines and Nonthaburi, Thailand, this chapter seeks to examine how developing cities can effectively plan and take specific measures to enhance climate and disaster-resilient cities.

8.2 Theoretical Framework and Methodology

8.2.1 Resilient City

Resilience is gaining increasing prominence within the literature on cities and climate change. There are many definitions of a resilient, ranging from very narrow to very broad and reflecting different theoretical and cultural values. It was first introduced to the field of ecology nearly four decades ago by Holling (1973):

"....resilience is the persistence of relationships within a system and the ability of these systems to absorb changes" (Holling 1973). Further, the urban ecological resilience literature (Berkes 2007; Folke 2006) defines resilience as the ability of a city or urban system to absorb disturbance while retaining identity, structure, and key process (Leichenko 2011). In other words, resilience is the capacity of a system to undergo disturbance and maintain its functions and controls (Gunderson and Holling 2001). Recently, the concept has also been applied to human social systems (Leichenko 2011). Resilience means the ability of a system, community, or society exposed to hazards to resist, absorb, accommodate to, and recover from the effects of a hazard in a timely and efficient manner, including through the preservation and restoration of its essential basic structures and functions (UNISDR 2010).

The UNDP defines resilience as an inherent as well as acquired condition achieved by managing risks over time at individual, household, community, and societal levels in ways that minimize costs, build capacity to manage and sustain development momentum, and maximize transformative potential (UN 2013). The United Kingdom Department for International Development's (DFID) definition also links resilience with long term development: "disaster Resilience is the ability of countries, communities and households to manage change, by maintaining or transforming living standards in the face of shocks or stresses – such as earthquakes, drought or violent conflict – without compromising their long-term prospects" (DFID 2011).

8.2.2 Planning Resilient Cities

As Table 8.1 shows, no single definition of resilience has yet to be adopted by all development and humanitarian actors. However, there appears to be general agreement that resilience is more than just the ability of a system (such as a household, community, or city) to bounce back to its pre-disaster state. Rather, it is a more integrated and holistic concept that focuses not only on a system's ability to withstand (absorb) shocks and stresses but also its ability to adapt to dynamic conditions and put in place mechanisms that enable longer term, systemic responses to the underlying causes of vulnerability (Barrett and Constas 2013).

Designing the resilience programs, the Department of International Development (DFID) of the Government of United Kingdom (UK) analyzes the levels of resilience by raising a question of how system will be reacted when they experienced a sudden shock or stress. Those who collapse or recover but are worse than before are likely to fall deeper into a vulnerability pathway. However, those who bounce back or bounce back better can be said to be on a resilience pathway. As Baba et al. reveals when risk exposure exceeds a certain level, effects on urban systems will begin to appear in a discontinuous manner. The measures that can be taken to this point include precautionary measures and adaptive measures to draw out the adaptability of the systems. Furthermore, when risk exposure surpasses a certain threshold, a regime shift (or revolutionary phenomenon) takes place, breaking down the existing framework of urban systems and demonstrating the transformative ability to create fundamentally new systems (Baba and Tanaka 2014).

Table 8.1 A selected definitions of resilience

Citation	Definition
Holling 1973	"…the capacity of an ecosystem to absorb shocks and maintaining functions"
Cadell et al. (2001)	"…the ability to adapt to, cope with and even be strengthened by adverse circumstances"
Ganor and Ben-Lavy (2003)	"The ability of…communities to deal with a state of continuous, long-term stress, which causes gaps between environmental stimuli and their functional coping behavior"
Frankenberger et al. (2007)	"…is built in a process of creating and strengthening personal, familial, social, organizational and economic systems to resist and cope effectively in times of stress, threats, crisis and emergencies"
DFID (2011)	"…the ability of …communities… to manage change, by maintaining or transforming the living standards in the face of shocks or stresses…without compromising their long-term prospects"
USAID (2012)	"…the ability of people, households, communities, countries, and systems to mitigate, adapt to, and recover from shocks and stresses in a manner that reduces chronic vulnerability and facilitates inclusive growth"
UNDP Drylands Development Centre (2013)	"…a transformative process of strengthening the capacity of… communities…to anticipate, prevent, recover, adapt and/or transform from shocks, stresses and change"
UNISDR (United Nations Strategy for Disaster Reduction) (2011)	"…..The ability of a system, community or society exposed to hazards to resist, absorb, accommodate to and recover from the effects of a hazard in a timely and efficient manner, including through the preservation and restoration of its essential basic structures and functions"
ICLEI – Local Governments for Sustainability (http://resilient-cities.iclei.org)	"……reduce vulnerability to extreme events and respond creatively to economic, social and environmental changes in order to increase their long-term sustainability"

Thus, building resilience requires an integrated approach, and a long-term commitment to improving three interlink capacities: absorptive capacity, adaptive capacity, and transformative capacity (Béné et al. 2012). *Absorptive capacity* is the ability to minimize exposure to shocks and stresses where possible and to recover quickly when exposed (Frankenberger et al. 2012). Improved disaster risk management is aimed at improving absorptive capacity at the community and household levels, helping them to both reduce disaster risk and absorb the impacts of shocks without suffering permanent, negative impacts on their longer term livelihood security.

Adaptive capacity involves making proactive and informed choices about alternative livelihood strategies based on changing conditions (Frankenberger et al. 2012). Interventions to improve adaptive capacity are aimed at improving the flexibility of households and communities to respond to longer term social, economic, and environmental changes. This necessarily entails promoting livelihood diversification, supporting asset accumulation, and improving the social and human capital available to vulnerable populations.

Transformative capacity relates to governance mechanisms, policies/regulations, infrastructure, community networks, and formal safety nets that are part of the wider system in which households and communities are embedded. Transformative capacity refers to system-level changes that enable more lasting resilience and often challenge the status quo in a substantial way (Béné et al. 2012). Each of these capacities is not mutually exclusive and exists at individual, household, community, state, and ecosystem levels.

Hypotheses about the most vulnerable populations and the primary constraints to their absorptive, adaptive, and transformative capacities were then used to develop theories of change that identify key leverage points ("domains of change") for enhancing resilience, and to illustrate the causal mechanisms through which project activities would lead to the achievement of strategic objectives. Theories of change are a critical outcome of a comprehensive risk analysis in that they allow for an iterative, adaptive, and nonlinear approach that is necessary for resilience programming. In the reminder of this chapter, we detail this challenge through the articulation of two case studies around the three following objectives: characterizing the city context, risks associated with natural disaster and climate change at the city level; identifying the most vulnerable segment (people, locations, and sectors) of the city; and assessing the city's ability to adapt to anticipated change in climate.

8.2.3 Methodology

The research methodology includes both literature review and case study analysis. The current literature linking cities and climate change, disaster risk reduction, climate change adaptation, and resilience planning was carried out through desk review and the web search. The literature review helped for identifying a theoretical framework for case study analysis. Participatory methods were utilized throughout the process of data collection and field facilitators were selected in each country to conduct the workshop with the help of city officials in the respective cities. A series of focus group discussions were conducted during the period of 2013–2014 with specific groups including citizens, city officials, city councilors, academic institutions, and community organizations to ensure that all the key groups are represented as much as possible to gain a better understanding regarding the risk context. A timeline method and risk matrix was elaborated with community members to identify the main hazards and prioritize them concerning their intensities and frequencies. Second, a detailed analysis of existing capacities was carried out and prioritized. As a next step, two city consultation workshops were organized (2014–2015) inviting different stakeholders (local leaders, nongovernment organizations, government agencies, the private sector, academic, and department heads) to make a resiliency plan identifying key barriers preventing the impact being reduced and priority measures (both short-term and long-term) to reduce the impact of these risks. These workshops were chaired by the Mayors of the respective cities ensuring political leadership for the implementation of the plans. In the following section, the case studies will be introduced and the main findings are summarized.

8.3 Case Study

8.3.1 Cebu City, Philippines

8.3.1.1 City Profile

Cebu City is the second largest growth center in the Philippines after Manila. It is located in the central eastern part of the Cebu province (Fig. 8.1). Cebu is the first Spanish settlement and the oldest city in the Philippines (Cebu City 2007). The total land area of the city is 291 km^2 which is divided into 80 barangays which is the smallest administrative division in the Philippines for the effective decentralization of the administrative functions and service provisions. Out of them, 50 are classified as urban and 30 as rural barangays. According to the Census of Population

Fig. 8.1 Location of the Cebu City

and Housing (2010), the total population of the city has been increased from 718,821 in 2000 to 866,171 in 2010 with 1.9% annual population growth (National Statistic Office 2012). Due to its strategic location near to the airport and seaport, Cebu is the fast growing center of commerce, trade, and service sectors in the region and a majority of its labor force (73%) is employed in these sectors (World Bank 2011).

However, Cebu City had to face many urban challenges as a fast growing city. About 30% of its households are occupied in informal settlements in the risky areas with lack of basic infrastructure, services, and the livelihoods (Cebu City, 2007). Cebu City also lacked adequate infrastructure to meet the current and future urban growth. For example, the energy demand is expected to grow at 4.5% annually since 2010 and the city needs to increase the electricity supply to meet the growing demand and also to reduce its dependency in outside sources. The existing transport, water supply, drainage, and solid waste management systems are also poor maintained and not adequate (World Bank 2011).

8.3.1.2 Risk Assessment

Cebu City has a long history of impacts from climate and natural hazards. The city is vulnerable to tropical storms and cyclones with their associated high winds and storms. Its topography is rugged and mountains with and elevation of 900 m above mean sea level. Plain lands are found on the shorelines occupying 8% of its total land area consisting 40 barangays and two-thirds of its total population. The rest of other 40 barangays are situated in the uplands. 28% of Cebu City is classified as 18% slope or less and 64% is classified as alienable and disposable. The rest is classified as forest and timberland. 76.3% of land is covered under the National Integrated Protected Areas System (NIPAS), while 23% of the 80 barangays are totally or partly located in the watershed areas. Steep slopes and deep soils cause landslides, and intense rainfall frequently causes severe floods. The migration to the city further pushes low-income families to build their houses in nonbuildable or high-risk areas. The participants who were part of the focus group discussions on risk assessment identified that floods, urban fires, and landslide as a priority list of risk of the city considering the frequency and its susceptibility (see Table 8.2).

8.3.1.3 Capacity of Disaster Risk Reduction and Climate Change Adaptation

At the national level, the Disaster Risk Reduction and Management (DRRM) Act of 2010 or Republic Act (RA) No. 10121 and its implementing rules and regulations supported by the Climate Change Act of 2009 (RA 9729) have significantly strengthened the institutional foundation for Disaster Risk Management (DRM) and Climate Change Adaptation (CCA) in the Philippines. The National Disaster Risk Reduction and Management Plan (NDRRMP), 2011–2028 provides the legal basis for

Table 8.2 A list of priority risks in Cebu City

Flooding	
Urban barangays (5 high, 21 moderate, and 30 low)	
Weather events: heavy rainfall and thunderstorms over a short period, and also predicted that climate change increased the risk of flooding	
Poor maintenance: faulty sewer networks, poor or insufficient drainage networks, inadequate maintenance of watercourses	
Development/planning issues: inappropriate development in flood plains, informal housing on land in a way that prevents rainfall from draining away naturally	
Fire	
Urban barangays (5 high, 21 moderate, and 30 low)	
Physical: poor planning, narrow roads, informal housing	
Human: Illegal electricity connections	
Landslide	
Mountain barangays (19 high, 13 moderate, 8 low)	
Geological causes: heavy rainfall, earthquakes	
Morphological causes: slope angle, vegetation change, erosion	
Human causes: excavation, land use change, mining, deforestation	

policies, plans, and programs to deal with major disasters. It shifts the national focus from a purely reactive approach to include risk management and preparedness and establishes links to CCA covering four thematic areas, such as (1) Disaster Prevention and Mitigation; (2) Disaster Preparedness; (3) Disaster Response; and (4) Disaster Rehabilitation and Recovery. By law, the Office of Civil Defense formulates and implements the NDRRMP and ensures that the physical framework, social, economic, and environmental plans of local and provincial governments are consistent with such plan. The local governments have a great responsibility in developing the system to effectively comply with the national legislations.

Based on these national regulations, the Mayor of Cebu City executed an Executive Order No. 12-9 in 2012 adopting the provision of RA 10121 to create the Cebu City Disaster Risk Reduction and Management Council (CCDRRMC) and establish an office for the same which is Cebu City Disaster Risk Reduction and Management Office (CCDRRMO). The CDRRMC is chaired by the Mayor or his nominee and includes the representatives from key city administrative departments (city planning, CDRRRMO, social welfare services, heath, city police, fire bureau, engineering and public works, public services, city administrator, and city budget officer), national/provincial officials (education, defense, science, and technology), private sector, and civil society organizations.

The CDRRMC is drafted a Cebu City Disaster Risk Reduction and Management Plan (2014–2016) outlining key measures for reducing and mitigating disaster risks enhancing disaster preparedness, responsive, recovery, and rehabilitation capacities. Under the plan, Cebu City has taken following key actions reinforcing priority areas:

- Together with the Mines and Geosciences Bureau (MGB) of the national government's Department of Environment and Natural Resources (DENR), the City has completed its geo-hazard mapping for all 80 Barangays under the City's territorial jurisdiction, analyzing potential hazards and evaluating existing conditions of vulnerability that together could potentially harm exposed people, property, services, livelihood, and the environment on which they depend (Fig. 8.2).
- Established a self-contained, secure, and impervious Command and Control Centre (C3) with communication infrastructure, which could generate relevant, real-time, and up-to-date information to enable individuals, communities, and organizations threatened by a hazard to prepare and to act appropriately and in sufficient time to reduce the possibility of harm or loss.
- To prepare the citizen for sudden disaster, CCDRRMO in partnership with barangays and NGOs continue to conduct seminar and workshops for citizens to increase their awareness and preparedness for the various natural hazards and teach them the suitable use of geo-hazard maps. Further, it organizes volunteers for medical responds, supply medical equipment and supplies. Development of Information Education Communication (IEC) materials; DRRM seminars & training, including Workshops on Comprehensive Risk, Evacuation & Resource Mapping in pilot barangays; Development of school curricula to include disaster risk reduction and management; Emergency drills in Barangays and schools (Fig. 8.3).

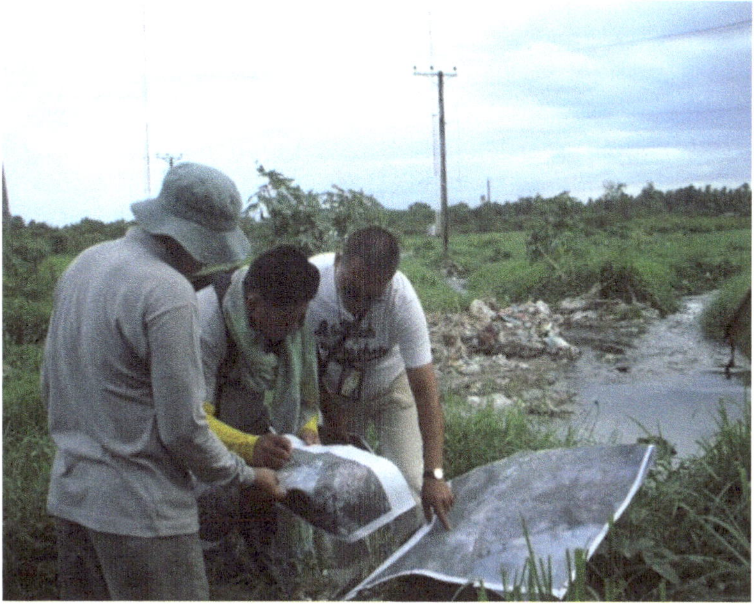

Fig. 8.2 Development of hazardous map by city officials

Fig. 8.3 Awareness program at barangays

Fig. 8.4 Distribution of tents for flood victims

- Further, city helps to return life to normal or improved the level of safety and security of the life and property as well called rehabilitation and recovery, Procurement of emergency equipment; Rapid assessment of damage and needs; Provisioning of tents, relief goods, and other supplies; and Management of evacuation centers. Relocation assistance; Provision of housing materials; Financial aid; and Crisis counseling (Fig. 8.4).
- Cebu City also pursues the policy of sustainable development, integrating disaster risk reduction and climate change adaptation into the city's development plans. Along this path, the City has started some initiatives like the Eco-Cities Network South East Asia led by the United Nations Industrial Development Organization (UNIDO), the APEC Low-Carbon Model Town, and the Climate and Clean Air Coalition to Reduce Short-Lived Climate Pollutants (CCAC).

8.3.1.4 Gaps and Lesson Learned

There are some institutional commitment and capacities to achieving DRR and CCA, but progress is not comprehensive or substantial. A lot of work still needs to be done in Cebu City to harmonize and mainstream DRR and CCA. The following gaps and lessons learned are identified during the discussions with city stakeholders:

Local Government's Accountability

The success or failure of DRRM plans, programs, and initiatives primarily lies in the hands of government officials who serve as the implementers of these initiatives. Participants of the workshop identified government officials from the national level down to the barangay level should accountable in mainstreaming DRRM in their agenda and plans. They pointed out that government officials must strictly implement laws and regulations related to DRRM and CCA. Participants said that we have good laws and regulations pertaining to environmental protection and disaster risk reduction and management, but what we lack are a clear framework or system where law offenders would be penalized and held accountable for violating such laws.

Eco-System Management

Cebu City is endowed with rich natural resources – the watersheds, the mangrove areas, the marine sanctuaries. However, recent disasters brought by natural hazards and man-made calamities have damaged thousands of lives, properties, infrastructure, and income opportunities. These threats to the increasing populace and growing businesses are getting more serious; thus a dire need to adopt concrete and stable solutions to address these issues. Part of the effort to mitigate natural disasters is environmental protection. The city should make effective measures to properly manage the forest, freshwater, and coastal eco-system.

Construction of Disaster-Resilient Infrastructures

The participants identified that one aspect the city could improve on in terms of DRRM is the construction of disaster-resilient infrastructures and buildings. The National Building Code of the Philippines provides this framework where it provides minimum standards and requirements in constructing and building private and public infrastructures and is consistent with the principles of environmental management and control. If strictly implemented and followed, this code could safeguard health, life, property, and public welfare, as well as reduce the potential impacts of disaster risks and climate change. However, the provisions of this code are only applicable to private and public infrastructures and do not apply to traditional indigenous family dwellings made of native materials such as bamboo, nipa, logs, and lumber. Dwellings made of light materials are the most vulnerable to disaster such as strong typhoons and fire incidence. Participants identified that family dwellings should be made of heavy materials which are more disaster-resilient and must not be located at disaster-prone areas.

Shifting to a Disaster-Resilient Mindset

According to the participants' responses, few of the hindering factors in achieving a disaster-resilient community are closely associated to sociocultural factors. These include social indifference from the community where those that are not directly affected are apathetic towards DRRM initiatives. The culture of silence, lack of political will from public officials, and corruption leading to substandard and superficial projects are also identified as sociocultural factors that hinder the success of DRRM plans and programs. According to the participants, this is one aspect that we should be working on. A shift of mindset and attitude is really important to achieve a disaster-resilient community. This is not achieved overnight and could take a longer process. But, if DRRM would be mainstreamed in Cebu City's agenda eventually, this would be inculcated in the society's culture and in the community's mindset.

Disaster Risk Reduction and Management Mainstreaming

There are multiple ways to mainstream DRRM and CCA in Cebu City. One of which is to include it in the plans and agenda of the city. Currently, Cebu City has a Disaster Management Plan. However, it does not lay down the concrete plans and next actions of Cebu City in terms of making the city resilient. Also, DRR plans are focused on response and recovery-related activities and do not really reflect action plans that are preparedness and prevention-related. A proposed solution for this is to formulate a more comprehensive Cebu City Action Plan and DRRM plans that are not only focused on responding to disaster but also on mitigating and preventing the impacts of disaster risk. Another problem of the current DRRM plans Cebu City has which was pointed out by the participants is that most are medium-term plans and term-based. Long-term DRR plans, according to participants, must be formulated so that there would be continuity of initiatives even after incumbent public officials would be replaced by another set of officials.

Education and Information Dissemination

Apart from formulating more comprehensive DRRM plans, another way to mainstream DRRM and CCA is through education campaign and information dissemination. One problem related to DRRM which was pointed out by the participants is the lack of awareness of the community regarding environmental laws and regulations including those related to DRRM. Most of the community members are aware that there are laws on DRRM and environment protection but are often not aware of what their provisions are. Seminars, trainings, and other information dissemination activities are good avenues to raise awareness within the community. As most hazards cannot be controlled, measures taken in advance of the impact of a hazard or stress are aimed at increasing the resilience of community to prevailing hazards. A multi-stakeholder consultation workshop was organized to identify the key barriers to resilience building and the points identified by the participants were summarized and grouped in the below figure.

8.3.2 Nonthaburi Municipality, Thailand

8.3.2.1 City Profile

Nonthaburi Municipality is the most urbanized center of the Nonthaburi Province, and it is very important city center in terms of administrative, business, and residential purposes. The city has a total land area of 38.9 km^2, comprised of 6 districts with 328 communities. Nonthaburi Municipality's population is approximately 257,742 people with an average population density of 6626 persons per square kilometer in 2014 (Nonthaburi City and IGES 2014). Nonthaburi Municipality is located on the eastern side of the Chao Praya River. Most of the city area is low plain. There are a lot of natural canals that were used in the past as a motor way which are currently functioning for drainage purpose (Fig. 8.5).

Nonthaburi city's climate is tropical wet, under the influence of the southwest monsoon. The economy of the municipality is mainly depended on the nonagricultural sector including industry service, wholesale and retail trade, hotel services, restaurants, etc. The total land area is almost urban consisting of residential area,

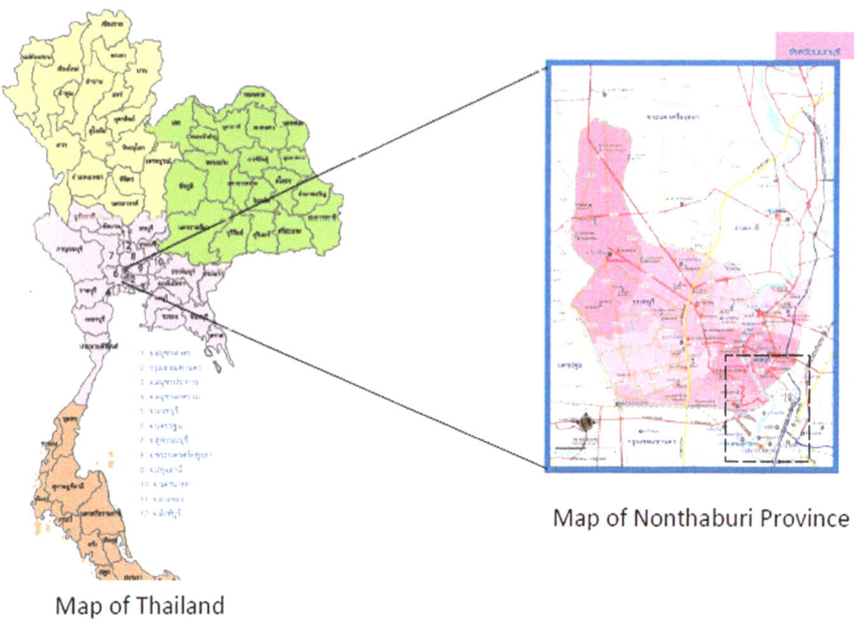

Fig. 8.5 Location of Nonthaburi Municipality

business zone, and government center such as the Ministry of Commerce, Ministry of Health, Department of Corrections, Army Quartermaster Department, and Government Lottery Office. Most of industrial enterprises in Nonthaburi Municipality are small-scale industries including wood products, food business, and business resource.

8.3.2.2 Risk Assessment

Flood is the most existing disaster risks in Nonthaburi Municipality. Historically, it has experienced flooding for several times including 1975, 1983, 1995, and 2005, but no floods can be compared to the Great Flood in 2011. It was a severe flooding for Bangkok and its outskirts in last 70 years since flood disaster in 1942. The Impact of flood has caused an enormous damage to life and property. Flood has destroyed the infrastructure, such as roads, buildings, etc. It has damaged agricultural crops, caused soil erosion, soil degradation, contamination of water, and killed many human lives. Since the end of July to November 2011, 65 of the country's 77 provinces were declared flood disaster zones, including Nonthaburi (Fig. 8.6).

There are two main factors for the severity of the flooding in 2011, such as natural causes and man-made factors. A natural characteristic of the afflicted regions and the effects from global warming caused the unexpected intensity of rainfall, which in turn caused flood inundation in the regions. The other cause was man-made, namely self-defense in flood prone areas, which caused isolation in flood protection, and the existing flood-management measures could handle only the average annual flooding (Kongchan, 2012). According to Nonthaburi, municipality area is located

Fig. 8.6 A major flood in 2011

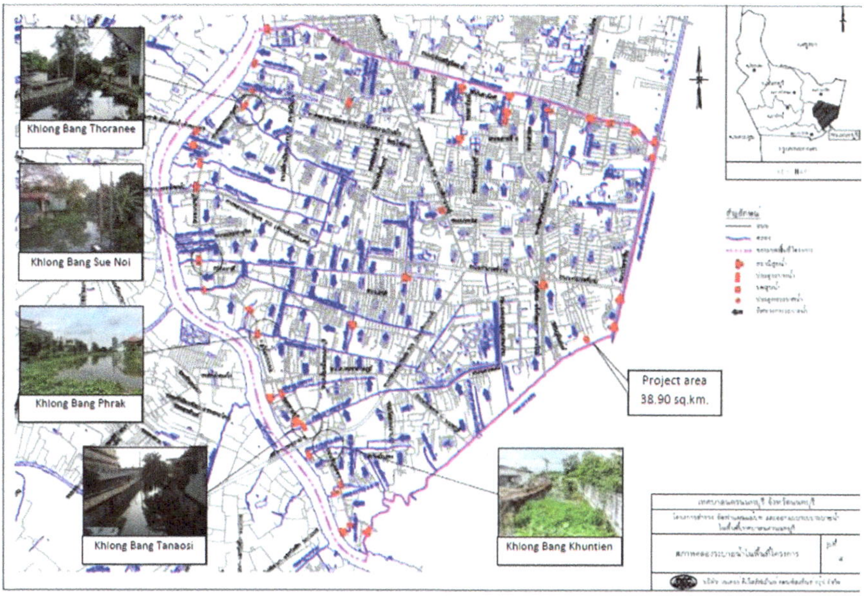

Fig. 8.7 Drainage system in Nonthaburi Municipality

on the flat, low-lying floodplain of Chao Pharya Basin with a height varying from +1.00 m MSL to +3.00 m MSL. The water is mostly drained to Chao Phraya river and surrounding major canals, e.g., Klong Bangkhen, Klong Bangtalad, and Klong Suay. However, when the water level in Chao Phraya River and major canals are already high, rain water drainage by gravity method or pump will be unsuccessful. This usually results in a stagnant flood situation, especially when there is a heavy rain. The areas that frequently suffer for stagnant flood are Chao Phraya bank. The municipality has its own flood and drainage system which have been built for a long period of time. Some of the canals or waterways are intruded by the building and structures. Including change of land use to dense community has caused itself a long-term flood problem in the area, especially the area located in Small Street where the sewerage systems are relatively small and uncompleted connection (Fig. 8.7).

8.3.2.3 Capacity of Disaster Risk Reduction and Climate Change Adaptation

The Disaster Prevention and Mitigation Act in 2007 gives the legal framework of Thailand's disaster management system. Following, the National Disaster Prevention and Mitigation Plan 2010–2014 was prepared by focusing on three main areas; (1) Principles of Disaster Management (2) Disaster Countermeasure Procedure and (3) Security Threat and Countermeasure Procedure in managing both natural disasters and national security issues. After the severe flood situation in Bangkok in 2011, the National Catastrophe Insurance Fund 2012 was established to provide risk management caused by floods, earthquake, and windstorm. The Department of Disaster Prevention and Mitigation (DDPM), under the Ministry of Interior, has been designated as the national focal point for disaster prevention, mitigation, and relief. The National Disaster Warning Center has established to enhance information sharing and earthquake and tsunami early warning in Thailand in cooperation agreements with several international organization such as Asian Disaster Preparedness Centre (ADPC), United States Agency for International Development (USAID), Indian Ocean Tsunami Warning System (IOTWS), and United States Trade and Development Agency (USTDA) in collaboration with Pacific Disaster Centre (PDC).

At the local government level, the preparation of local development plans is considered the first priority according to the municipal authority as well as the Determining Plans and Process of Decentralization to Local Government Organization Act B.E. 2542 (1999). Thus, local governments are prepared for changes and possible impacts that may occur. These local development plans are also in line with the regulations of the Ministry of Interior on the preparation and coordination of the development plan of the local government B.E. 2548 (2005), as well as the procedure for local government budget B.E.2541 (1998), Article 22; use of local development plans as a guideline for budgeting, etc. Furthermore, the local development plans must conform to the National Economic and Social Development Plan, national agenda, government policies, provincial development strategy and plan, policies of the local administrators, and problems and needs of people (community plan) (Fig. 8.8).

However, mainstreaming disaster risk assessment and climate change adaptation into the local development plans often lack. Nonthaburi Municipality has taken efforts to establish the Flood Prevention and Mitigation Centre in 2011 under the Municipal Order No 1069/2554 soon after the great flood in 2011. This center is headed by the Municipal Mayor and functions as a main center in coordinating,

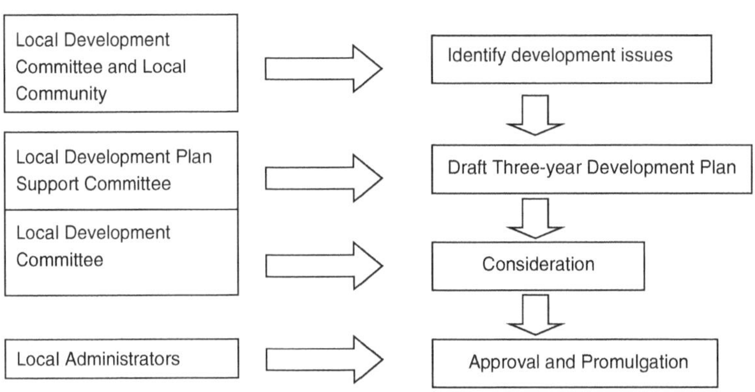

Fig. 8.8 Process of preparation of development plans

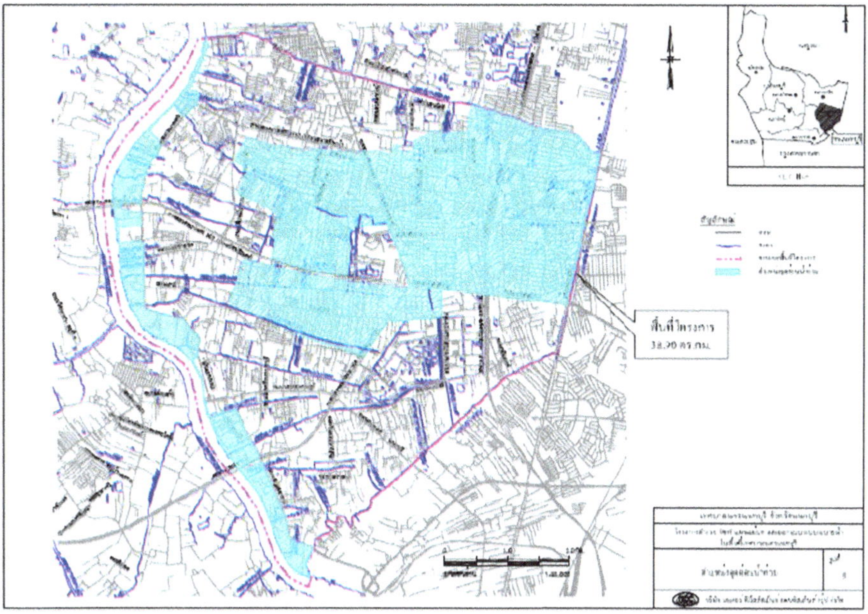

Fig. 8.9 Flood management area

monitoring, and planning flood prevention measures in the city. Nonthaburi Municipality has formulated a flood prevention guideline to prepare for future disasters. For the past 3 years, Nonthaburi Municipality has expedited several plans and flood prevention projects, focusing on the construction and maintenance of drainage systems (Fig. 8.9). There is also an emphasis on prevention measure including a preliminary design of flood prevention from river, a preliminary design of flood prevention from rain water, and construction of flood prevention structures, in total 47 projects, approximately 388,675,730 THB.

8.3.2.4 Gaps and Lessons Learned

Development of innovative strategies to achieve urban flood-resilient city is one of the priorities in Nonthaburi Municipality especially after the great flood in 2011. The municipality has developed some innovative strategies to enhance the resilience of cities to floods which include the following.

Transition from Flood Control to Flood Management in the City

Since 2011, the Nonthaburi Municipality has given more priority to flood disaster prevention and reduction. On the one hand, structural measures such as dyke reinforcement, river regulation, the construction of reservoirs, and the building of flood retention and storage basins have been reinforced. On the other hand, nonstructural measures such as decision support systems for flood prevention, the development of flood control operation pre-schemes, the management of flood storage and retarding basins, and flood insurance have been strengthened as well. Land use planning in municipal and nearby area; land use control and monitoring; specify and announce the flood risk area; create the flood risk map for warning and rescuing purpose; formulate flood warning system; prepare flood relief operation plan that is functional during the real situation; prepare flood restoration plan to be used after flood situation; promote the public relation campaign to raise awareness on flood prevention.

Community Ownership and Participation

In the case of disaster that has extensive coverage with many localities affected, the communities could not expect help from the government since the government has to distribute aids to many areas. Sometimes, the help arrived after the incident, or was insufficient, or even failed to be delivered. Hence, the local communities need to mobilize their own resource as much as possible including a preparation of human resource, a skillfulness and understanding of their area, self-awareness in their strength and weakness, and advance planning and project. For example, in case of flood prevention, local communities can conduct a survey and create general information of the area as well as the prevention measures. This will help protect or mitigate the magnitude of plausible disaster. However, there are some elements that need to be considered: (1) The capability of local people in initiating and sustaining their own community development; (2) Local leadership; (3) Broad-based local participation in comprehensive planning; (4) Education opportunities; (5) The improved utilization of local resources; (6) Responsible utilization of outside financial support; (7) Local initiative continuity; (8) The realization that responsibility for change rests with community members; and (9) Variation of perceptions of risks among community members and groups in community (ADPC 2006).

The Monitoring and Warning System in Chao Phraya Basin and Gulf of Thailand

To forecast the flow from upstream of Chao Phraya River, the municipality should closely coordinate with responsible organizations, including the Royal Irrigation Department to obtain the forecast water level data in advance. The daily forecast sea level data from the Naval Hydrographic Department and the daily observed tide data from Royal Thai Navy should be used to check with the current flood prevention system in order to be well prepared in advance, especially during the wet season which Chao Phraya will receive water from upstream. The daily and weekly forecast rainfall data from the Thai Meteorological Department will be useful for efficient water management in the area. The immediate benefits of a flood early warning system are the reduction in the loss of human lives and the mitigation of damage to properties. More importantly, it allows more direct and quicker community access to the results of warning devices. This enhances the community's preparedness for flood events. It is envisioned that the local government units will be able to use the tool to derive flood indices or establish the correlation between rainfall intensity and the resulting height and extent of inundation in their area.

Mainstreaming Flood Management Strategies in to the Local Development Plans

Flood management strategies need to be an integral part of a municipality's development plans. The current development plan does not directly address disaster risk reduction but incorporates it in the areas of social welfare, water resource management, agriculture, and rural development. Within these priority sectors, the disaster risk reduction activities identified include protecting rural areas from natural hazards, enabling communities for disaster preparedness and risk reduction, and reducing the vulnerability of the poor to external shocks, including natural hazards.

8.4 Building Resilient Cities: Critical Observations and Discussions

The case study analysis identified several observations particularly important for understanding the challenges and opportunities for building resilient cities in developing countries in Asia. These include the following.

8.4.1 Comprehensive Risk Assessment and Linking Them to Decision Making

Information is crucial to address resilience planning. Good program design depends on comprehensive analysis of all the key factors that affect the system(s) under study. Analysis begins with a comprehensive understanding of the environmental, political, social, economic, historical, demographic, religious, conflict, and policy conditions that affect, and are affected by, how households, communities, and governments prevent, cope with, and recover from shocks and stresses. A comprehensive assessment is necessary to fully understand the constantly changing relationship between risk and vulnerability on the one hand and livelihood outcomes and resilience capacities on the other hand.

Unfortunately, a readily available data on climate risks and hazards (trends and projections) do not exist in the cities. Carrying out a comprehensive risk analyses and data gathering are also costly and local government staff lack scientific background and expertise. The community-based, participatory risk analysis adopted by the case study cities shows great potential in understanding the context better and reduce both costs and time for data collection. The participatory consultative method will engage city experts and stakeholders in each city in the risk and capacity assessment process also help to develop local adaptive capacity. The qualitative data gathered from participatory methods can integrate with scientific data of national and global disaster and climate models providing a deeper understanding of risks and the scientific foundation for planning efforts by city-decision makers and other stakeholder groups. Risk assessments can help governments make better-informed decisions. Systematic screening of risks can also help determine the level of risks to people and assets and guide options for risk management.

8.4.2 Integrated and Long-Term Approaches

Resilience building relies on integrated programming – a cross-sectoral approach with a long-term commitment to improving the three critical capacities: absorptive capacity (disaster risk management), adaptive capacity (longer-term climate change adaptation), and transformative capacity (improved governance and enabling conditions). Currently, both case study cities are focusing more on actions for building absorptive capacity with adopting DRR measures, including risk assessment and preparedness planning, disaster response, relief and recovery programs. Both cities have established disaster management committees, early warning system and command centers, and information and education campaign for its citizens to well prepare and response to disasters. Also they have well-established organizational mechanisms at community level for sudden response (volunteers and training) and emergency equipment.

However, both city governments have begun to recognize the importance of strengthening adaptive capacity under the resilience capacity-building initiatives to address the current threats and those emerging from climate change. The built and ecological environment of cities significantly makes climate change impacts. For example, the presence of lack of street trees and parks, coverage of flood retention areas, extent of wastewater and drainage system, waste disposal to drainage and encroachment, and illegal construction can intensify flooding in the cities. For responding to these issues, local governments identified both structural and non-structural initiatives that can be pursued through legal and political system, zoning regulations, infrastructure, and urban services specifying change building codes and restrict land use in areas subject to climate change impacts such as flooding, increase urban tree coverage and vegetation, plan and manage the river-basin and upper land's vegetation, cleaning of the channels, and construction of new drainage system if necessary.

Further, this study found that both city governments face many challenges in their efforts to planning resilient cities due to fiscal and policy-making limitations.

In order to effectively address the challenges faced by the cities in integrating resilient plans into daily decision making and long-term development plans of the cities, emphasis was given on enhancing the transformative capacities, which would be needed in order to facilitate systemic changes in the structural constraints (such as those of ecological, political, economic, or social structures). To bring typical changes into local policies, institutional and administrative process, and planning systems where the changes are often most needed, four key factors are identified, such as effective political leadership, efficient financing, jurisdictional coordination, and citizen participation.

8.4.3 Collaborative Partnerships

Resilient building requires cooperation on different sectors and levels. Effective implementation of resilient policies needs cooperation between local stakeholders, such as authorities, businesses, and citizens. Many city governments are weakened due to only limited power over and responsibility for key public services such as planning, land use management, transportation, water, and sanitation systems. Existing governing institutions are also often horizontally fragmented, uncoordinated, and in many cases ad hoc. Coordination is fundamental not only in basic sectoral areas such as land, transport, energy, emergency preparedness, but in addressing city, regional, and national levels of climate governance.

8.5 Conclusion and Recommendations

This study identified that local governments have great potential to lead on building resilient cities, despite the political, technical, and financial constraints that they are faced in day-to-day operation. In order to effectively address the challenges of natural and climate change disasters, cities need to incorporate risk assessment and resilient measures into daily decision making and long-term development plans and investments. Following are some recommendations for building resilience capacity in developing cities:

Risk-Informed Program Design Effective interventions for addressing resilience require well-designed programs based on a theory of change that correctly identifies appropriate leverage points needed to effect desired outcomes. Good program design for building resilience capacity requires a comprehensive multihazard, multisector assessment of all the contextual factors that affect the system(s) under study, which then informs the theory of change. A comprehensive assessment is necessary to fully understand the constantly changing relationship between risk and vulnerability on the one hand and livelihood outcomes and resilience on the other hand. The participatory, community-based methods can link with the scientific, quantitative data gathered from other sources. More innovative donor funding mechanisms are needed in order to support local governments for building capacity to conduct comprehensive risk analysis, develop meaningful theories of change, and design appropriate interventions to address underlying causes of vulnerability and risk.

Long-Term, Integrated Approaches to Resilience Programming A cross-sectoral approach with a long-term commitment is required in order to improve the absorptive, adaptive, and transformative capacities of vulnerable populations to shocks and stresses. Programs with an integrated approach for improving cross-sectoral outcomes ensure that partners and sectors work together to address key leverage points and adopt complementary, synergistic strategies to promote resilience. Local governments need develop both short-term and long-term resilience plans, but many more need to bring them into their everyday operations.

Strategic Collaboration to Enhance Transformative Capacity Local governments face many challenges in their efforts to building resilient cities due to lack of transformative capacity. Collaborative efforts, alliances, or high-level task forces that involve donors, UN agencies, and national governments can more effectively improve transformative capacity at local level, greatly enhancing local government's initiatives to improve the resilience in their respective cities.

References

ADPC (2006) Participant's workbook: community-based disaster risk management for local authorities, Bangkok. http://www.unisdr.org/files/3366_3366CBDRMShesh.pdf. Accessed 20 June 2019

Baba K, Tanaka M (2014) Development of a concept of resilient city and indicators for resilient assessment, 1st conference of International Alliance for Sustainable Urbanization and Regenaration, Kashiwa-no-ha, Japan

Barrett CB, Constas M (2013) Toward a theory of resilience for international development applications. Unpublished working paper. Cornell University, Ithaca

Béné C, Wood RG, Newsham A, Davies M (2012) Resilience: new utopia or new tyranny? Reflection about the potentials and limits of the concept of resilience in relation to vulnerability reduction programmes, IDS Working Paper, vol. 2012, no. 405

Berkes F (2007) Understanding uncertainty and reducing vulnerability: lessons from resilience thinking. Nat Hazards 41:283–295

Cadell S, Karabanow J, Sanchez M (2001) Communiy empowrment and resilience: paths to wellness. Can J Commun Ment Health 20:21–36

DFID (2011) Defining disaster resilience – a DFID approach paper, London

Folke C (2006) Resilience: the emergence of a perspective for social-ecological systems analyses. Glob Environ Chang 16:253–267

Frankenberger TR, Sutter P, Teshome A, Aberra A, Tefera M, Tefera A, Taffesse A, Ejigsemahu Y (2007) Ethiopia: the path of self-resiliency, Vol 1. Final Report. http://www.cangoethiopia.org/asset/docs/(2)%20Self-Resiliency%20Report%20-%20CANG)-CHF%20(08.08.07).pdf. Accessed 20 June 2019

Frankenberger TR, Langworthy M, Spangler T, Nelson S (2012) Enhancing resilience to food security shocks in Africa. Discussion paper. http://www.fsnnetwork.org/sites/default/files/discussion_paper_usaid_dfid_wb_nov._8_2012.pdf. Accessed by 20 June 2019

Ganor M, Ben-Lavy Y (2003) Community resilience: lessons derived from Gilo under fire. J Jew Communal Serv:105–109

Gunderson L, Holling CS (eds) (2001) Panarchy: understanding transformations in human and natural systems. Island Press, Washington, DC

Holling C (1973) Resilience and stability of ecological systems. Annu Rev Ecol Syst 4:1–23

http://www.unisdr.org/campaign/resilientcities/about. Accessed 29 June 2015

IPCC (2007) Climate change 2007: impact, adaptation and vulnerability, contribution of Working Group II to the fourth assessment report of the Intergovernmental Panel on Climate Change. Cambridge University Press, Cambridge

IPCC (2012) Managing the risks of extreme events and disasters to advance climate change adaptation, a special report of Working Groups I and II of the Intergovernmental Panel on Climate Change. Cambridge University Press, Cambridge

Leichenko R (2011) Climate change and urban resilience. Curr Opin Environ Sustain 3(3):164–168

Mitlin D, Satterthwaite D (2013) Urban poverty in the global south: scale and nature. Routledge, Abingdon/Oxon/New York

National Statistic Office (2012) The 2010 census of population and housing reveals the Philippine at 92.34 Million. http://www.psa.gov.ph/content/2010-census-population-and-housing-reveals-philippine-population-9234-million. Accessed 20 June 2019

Nonthaburi City and IGES (2014) Report on development of resilient city strategy for flood prevention in Nonthaburi City. IGES, Kitakyushu

Pelling M (2003) The vulnerability of cities. Earthscan, London

Prabhakar SVRK, Sirinivasan A, Shaw R (2009) Climate change and local level disaster risk reduction planning: need, opportunities and challenges. Mitig Adapt Strateg Glob Chang 14:7–13

Quote by Ban Ki-moon, UN Secretary-General at the Incheon Conference "Building an Alliance of Local Governments for Disaster Risk Reduction", August 2009

UCCRN (2011) Climate change and cities – first assessment report of the urban climate change research network, urban climate change research network. Cambridge University Press, Cambridge

UNDP Drylands Development Centre (2013) Community-based resilience assessment (CoBRA) conceptual framework and methodlogy. http://www.disasterriskreduction.net/fileadmin/user_upload/drought/docs/CoBRA%20Conceptual%20Framework%20and%20Methodology%20-%20Post%20Arusha%20-%204%20April%202013.pdf. Accessed 20 June 2019

UNISDR (2010) Making cities resilient: my city is getting ready, 2010–2011. World disaster reduction campaign. UNISDR, Geneva

United State Agency for International Development (USAID) (2012) Building resilience to recurrent crisis: USAID policy and programme guidance. http://www.usaid.gov/sites/default/files/documents/1870/USAIDResiliencePolicyGuidanceDocument.pdf. Accessed 20 June 2019

UNISDR (2011) Global assessment report on disaster risk reduction: revealing risk, redefining development. UNISDR/Practical Action, Geneva

UNISDR (2015) Sendai framework for disaster risk reduction 2015–2030. http://www.wcdrr.org/preparatory/post2015. Accessed by 29 June 2015

United Nations (2013) A new global partnership: eradicate poverty and transform economies through sustainable development the report of the high-level panel of eminent persons on the post-2015 development agenda. United Nations Publications, New York

WMO (2008) Future climate change research and observations: GCOS, WCRP and IGBP learning from the IPCC fourth assessment report. Workshop and Survey Report, IGBP Report No. 58 (WMO/TD No. 1418)

World Bank (2011) Climate change, disaster risks and the urban poor: cities building resilience for a changing world. World Bank, Washington, DC

World Bank and GFDRR (2013) Building resilience – integrating climate and disaster risk into development. World Bank, Washington, DC

Chapter 9
Strengthening Urban Resilience/Disaster Risk Management in Asian Cities

Tadashi Matsumoto

Abstract This chapter briefly presents OECD's recent study which has assessed disaster risk management policies to enhance urban resilience in Southeast Asian cities. Between 2014 and 2016, the OECD analyzed disaster risk management policies of five Southeast Asian cities: Bandung (Indonesia), Bangkok (Thailand), Cebu (Philippines), Hai Phong (Viet Nam), and Iskandar (Malaysia). All the five cities face significant natural disasters risks as other Southeast Asian cities, but with substantial difference in the urban policy contexts. Based on the analysis, the study provides overall assessment and recommendations of disaster risk management policies for Southeast Asian cities, as well as individual assessment and recommendations for each case city. Among the OECD findings, this chapter highlights the need for integrating disaster risk management into urban green growth policy frameworks, as well as the need for metropolitan coordination. For example, the case of Bangkok clearly demonstrates the need for aligning "hard" investments in flood-resilient urban infrastructure with land use planning, zoning policies as well as with "soft" (i.e., nonstructural) instruments at the metropolitan scale. The chapter also underlines that successful disaster risk management outcomes will require locally tailored policies, taking different urban policy contexts into account.

Keywords OECD · Urban green growth · Urban resilience · Disaster risk management · Bangkok · Urban infrastructure · Soft resilience measures

9.1 Introduction

Asian cities are particularly vulnerable to, but not well prepared to cope with, risks related to natural disasters, in particular flooding due to a rise in sea level, rapid urbanization without adequate urban infrastructure, and natural and human-induced subsidence. According to Munich RE, between 1980 and 2017, Asia suffered

T. Matsumoto (✉)
OECD, Paris Cedex 16, France
e-mail: Tadashi.MATSUMOTO@oecd.org

M. Tanaka, K. Baba (eds.), *Resilient Policies in Asian Cities*,
https://doi.org/10.1007/978-981-13-8600-8_9

disproportionately from natural disasters, both in terms of lives lost (over 1.2 million people or 71% total global loss of life) and uninsured assets (89% of total global losses amounting to USD 1.69 trillion). Such disasters can severely affect urban dwellers, the local environment, and long-term economic growth. People living in slums and informal settlements, in particular, should be privileged targets of risk management policies. They often live in hazardous locations such as flood plains, steep hillsides and low-lying coastal areas without proper protection against floods, proper drainage, waste removal, and roads. The consequences of a major flood or storm can be devastating in such areas and can slow urban development and the pursuit of social equity considerably. Recognizing the complexity of uncertainty associated with climate change predictions, urban climate change resilience (UCCR) is becoming a key concept to consider (Matsumoto and Daudey 2014).

Asian cities have experienced one of the most dynamic urbanization processes in the world. The combined urbanization level of the ASEAN-5 countries – Malaysia, Thailand, Indonesia, the Philippines, and Viet Nam – increased from 29.5% in 1980 to 51.4% in 2015 and is expected to reach 67.7% in 2050 (UN DESA 2014). In parallel, gross domestic product (GDP) per capita in the ASEAN-5 region increased from around USD 8,500 in 1985 to around USD 24,800 in 2015, despite economic crises that hit the Asian economy in 1997 and in 2008 (OECD 2016). However, this positive growth is marred by poor environmental performance, widening social disparities, and increasing exposure and vulnerability to natural disasters. Asia urgently needs to shift its growth model to preserve natural assets and environmental services, and promote social development, which will ensure long-term sustainable growth. The concept of *urban green growth* in this regard can be a powerful vector of sustainable and resilient cities by emphasizing the existence and potential of co-benefits between economic and environmental performance (OECD 2016).

In 2014, the OECD started a study: *An Assessment of Disaster Risk Management Policies in Southeast Asia*, as a part of the OECD project: *Urban Green Growth in Dynamic Asia.*[1] The study was supported by the Global Initiative on Disaster Risk Management (GIDRM)[2] and OECD Knowledge Sharing Alliance (KSA).[3] After

[1] The Urban Green Growth in Dynamic Asia project is a research project conducted by the OECD in 2013–2018. It explored how to promote green growth in cities in Asia, examining policies and governance practices that encourage environmental sustainability and competitiveness in a rapidly expanding economy. The main aim was to assist Southeast Asian cities in decoupling economic growth from environmental stress, and to promote a long-term trajectory of sustained growth.

[2] GIDRM is a project commissioned by the German Federal Ministry of Economic Cooperation and Development (BMZ) to the *Deutsche Gesellschaft für Internationale Zusammenarbeit* (GIZ) GmbH, aiming to strengthen German's contribution to improved disaster risk management worldwide. The initiative has three priority areas: (1) strengthening disaster response preparedness and civil protection; (2) resilient critical infrastructure and economic cycles; and (3) effective early warning systems.

[3] KSA was created in January 2013, in partnership with BMZ and Korean Ministry for Strategy and Finance. It aims to leverage the OECD's multidisciplinary expertise, engage in conversations and mutual learning processes with emerging and developing economies, and increase impact by working closely with multi- and bilateral organizations having field presence and implementation capacities and/or networks. The Urban Green Growth in Dynamic Asia project was one of the pilot projects of KSA.

3 years of work, the final report was recently published (OECD 2018). This chapter briefly presents the report, highlighting the assessment framework for disaster risk management policies (Sect. 9.2), the overall assessment and policy recommendations (Sect. 9.3), and the case of Bangkok Metropolitan Region, Thailand (Sect. 9.4).

9.2 Assessment Framework for Disaster Risk Management Policies

The OECD study – *An Assessment of Disaster Risk Management Policies in Southeast Asia* – aimed at analyzing national and subnational disaster risk management policies to enhance urban resilience. More specifically, the objectives of the study were following (OECD 2018):

- To identify the variety of disaster risk management challenges under different geographical, socioeconomic, and environmental contexts of Asian cities
- To overview ongoing policy practices to promote disaster risk management and assess their impacts
- To propose more efficient and effective policy options to policymakers and implementation partners in Asia.

The assessment was based on the argument that the following three steps are required to institute effective disaster risk management. These steps will help Southeast Asian cities to better manage the worst disaster risks when they strike, thus making them more resilient (OECD 2018):

1. Preparation: Preparation incorporates the crucial early stages of hazard identification and strategic policy formation. A lack of preparation exacerbates the risks posed by environmental hazards that may otherwise be avoided because it does not set in motion more advanced disaster risk management practices. Preparation provides a foundation to manage and limit damage, also considering the cost and time required to recover from a natural disaster. Effective preparation should also identify governance challenges and potential cross-sectoral coordination partners, in addition to generating public awareness.
2. Prevention: Prevention should be proactive and reduce exposure to disaster risks in the long-term through regulation, fiscal instruments, as well as investment in resilient urban infrastructure. For instance, risk-sensitive land use could direct urban development to areas that will not be exposed to regular flooding, such as floodplains.
3. Response: Response should facilitate a city's recovery from natural disasters. To be most effective, careful policy coordination and planning should occur in advance to "build back better" rather than lapsing into a process known as "reconstructing vulnerability." Insurance and contingency funds are fundamental

components of effective response planning, to prepare for unexpected costs arising from natural disasters. Both the public and private sector should contribute to the design and implementation of policies to enhance disaster response. For example, while insurance is normally provided by the private sector, national and local governments can create market conditions and provide contingency lines of credit to complement private insurance in areas where insurance would be socially desirable but is not provided by the private sector.

The study utilized the OECD's definition of resilience as "the ability of households, communities and nations to absorb and recover from shocks, whilst positively adapting and transforming their structures and means for living in the face of long-term stresses, change and uncertainty" (OECD 2013). It also used the IPCC's disaster risk framework as the starting point for its assessment (IPCC 2012), but focused less on the risks themselves and rather on the institutional frameworks, governance structures, and policy approaches affecting each city's disaster risk management strategy to enhance urban resilience (OECD 2018).

The study included five city cases from multiple countries in Southeast Asia, in order to reflect the variations of geographical, socioeconomic, and environmental contexts. The five cities were: Bandung (Indonesia), Bangkok (Thailand), Cebu (Philippines), Hai Phong (Viet Nam), and Iskandar (Malaysia). All the five cities face significant natural disasters risks as other Southeast Asian cities, but with substantial difference in the urban policy contexts (Table 9.1).

Each case study analyzed the following elements, based on the information received through a questionnaire and through meetings and interviews to key actors (policymakers from local/regional/national governments, academics and researchers, business community, NGOs, etc.) (OECD 2018):

Table 9.1 Overview of the five case study cities

Name	Population (million)	Annual population growth	Area (km^2)	Density (people/km^2)	Gross regional product per capita (PPP)
Bandung (Bandung metropolitan area)	8.6 (2015)	1.94% (2000–2010)	3509	2452	USD 7,490 (IDR 10.69 million) in 2012 (constant 2000 prices)
Bangkok (Bangkok metropolitan region)	14.5 (2010)	0.9% (2004–2014)	7762	1347	USD 29,540 (THB 359,798) in 2012 (constant 2005 prices)
Cebu (metro Cebu)	2.8 (2015)	2.9% (2000–2010)	1163	2450	USD 5,084 (PHP 69,700) in 2012 (constant 2000 prices)
Hai Phong (city)	2.0 (2015)	1.0% (2000–2010)	1527	1284	USD 3,940 (VND 22 million) in 2015
Iskandar (Iskandar Malaysia economic region)	2.0 (Est. 2015)	3.7% (2005–2015)	2300	870	USD 19,262 (RM 27,631) in 2013 (constant 2005 prices)

Source: OECD (2016)

- Economic and environmental trends relevant to disaster risk management, including of the assessment of natural and man-made disaster risks
- Key plans and strategies to promote disaster risk management (key metropolitan-wide strategies, plans, or visions to address local conditions; key nation-wide strategies, plans, or visions relevant to urban resilience disaster risk management at the city level)
- Assessment of policies to promote disaster risk management (policies currently undertaken by various authorities; identifying gaps between the trends and the policy responses; assessment of the effectiveness of policies; options for improvement)
- Enabling strategies to promote disaster risk management (the connections, mechanisms, and instruments at play between the metropolitan and national strategies/plans/visions; financing urban resilience/disaster risk management actions; diversity of the policy instruments used by the local government; horizontal cooperation and community engagement; capacity building and knowledge sharing)

In addition to the questionnaire and interviews, five knowledge-sharing workshops were organized under the framework of the OECD *Urban Green Growth in Dynamic Asia* project in 2014–2015. These were particularly useful platforms for exchanging knowledge and sharing policy practices, among OECD countries, fast-growing Asian countries, international organizations, and research institutions.

9.3 Key Findings: Integrating Disaster Risk Management into Urban Green Growth Policy Framework

The overall assessment of disaster risk management policies in Southeast Asian cities, based on the five case studies, are summarized as follows (OECD 2018):

- Preparedness: Southeast Asian cities are largely underprepared for natural disaster risks, especially as regards vulnerability and risk assessment practices. Comprehensive hazard assessment and mapping is not uniformly employed, which is particularly harmful for identifying and protecting low-income communities at risk.
- Land use: Land use policies do not often consider disaster risk management, which has resulted in continued urban development in risk-prone areas.
- Urban infrastructure: Two-thirds of Asia's infrastructure needs by 2050 still have to be built and financed, thus providing an opportunity to factor in resilience to natural disasters. The large need for infrastructure investment will require large-scale private sector engagement. To this end, public finance plays a critical role to facilitate, leverage, and guide private investment. At the city level, this is a challenge when tax revenues collected by local governments are often small.

- Insurance: Adequate private and public insurance mechanisms to share disasters risks are not well developed in the case study cities. Almost three quarters of all financial damages globally are not insured, and this insurance gap is even more pronounced in Asia.
- Governance: The coordination mechanisms between national and local governments are often lacking or not clearly defined, obstructing the implementation of national disaster risk management policy frameworks (when they exist) at the local level.
- Stakeholder engagement: While engaging local communities from the early stages of decision making can help develop more effective and inclusive disaster risk management strategies and frameworks, such opportunities are not always offered in the case study cities.

Based on these assessments, the OECD study recommends the following disaster risk management actions to enhance urban resilience in Southeast Asian cities (OECD 2018):

- Conduct a comprehensive vulnerability and risk assessment to develop a local resilience action plan.
- Adopt risk-sensitive land use policies combining regulatory and fiscal instruments to guide urban development away from risk-prone areas.
- Integrate disaster risk management policies and urban green growth policies, especially in the infrastructure sector, to generate "co-benefits."
- Develop disaster risk financing mechanisms to serve as a backbone of effective disaster response planning.
- Promote the use of information and communication technologies.
- Foster vertical and horizontal coordination to foster a "whole-of-government" approach.
- Engage with stakeholders to promote inclusiveness and encourage a culture of disaster risk management.

Among these assessments and recommendations, this chapter highlights the need for integrating disaster risk management into urban green growth policy framework. The OECD study has demonstrated that policy complementarities and synergies are often observed between disaster risk management and other policy objectives related to the natural environment. For example, ecosystem-based adaptation measures are presented as a considerable, yet unrealized, investment opportunity to enhance urban resilience (OECD 2018):

- *Rainwater harvesting* presents an interesting opportunity in Southeast Asian cities. In the case of Bandung, Indonesia, it has been incorporated into Bandung's Integrated Rainwater and Wastewater Management System and is expected to reduce pressure on the city's drainage infrastructure and diminish the risk of flooding (Lee 2013).
- *Swales* and *green curbs*, which can increase natural vegetation in cities, create pedestrian space and de-incentivize private car use while also serving their function of retaining rainwater run-off to minimize flash flooding (in comparison to

more conventional polder or drainage infrastructure). In addition to offering increased absorptive capacity during heavy rainfall, swales and green curbs are also likely to win public support more easily and be implemented more quickly than other large and disruptive solutions.

Such measures, while they can also generate benefits for urban green growth (e.g., higher urban quality of life), are often more cost-effective than conventional large-scale urban infrastructure investment.

Another important dimension of the OECD's assessment and recommendations is the need for metropolitan coordination. It stresses that land use, water resources management, and flood risk are typical, interconnected urban resilience and resource efficiency challenges that bear a metropolitan dimension. However, the lack of coordinating mechanisms at the metropolitan scale often presents obstacles to urban resilience. For example, the development of the Bus Rapid Transit which is only being designed and implemented within Cebu City, despite metropolitan commuting flows, may create expensive infrastructure lock-in (and rising GHG emissions) that will be difficult to circumvent in the future (Daudey and Matsumoto 2017).

The study argues that national governments should therefore spearhead metropolitan governance initiatives and allocate sufficient financial and political resources for their proper functioning. Like in many OECD metropolitan areas, there is an opportunity for greater collaboration between local governments at the metropolitan level in Southeast Asian cities. Since 2007, Iskandar has benefited from the establishment of the Iskandar Regional Development Authority (IRDA), to coordinate various public agencies and private stakeholders in the metropolitan area. In contrast, in the Bangkok Metropolitan Region (BMR), no such authority exists and coordination mechanisms at the metropolitan scale are lacking (OECD 2015, 2018).

The study also points out that *functional urban areas*, as defined by the European Commission and the OECD (2012), would be a useful concept as a unit of policy coordination. A functional urban area typically goes beyond the administrative border of a city's core municipality and includes other surrounding districts of an urban area, which would lead to more coherent metropolitan disaster risk management policies and strategies and plans (OECD 2018).

9.4 The Case of Bangkok

The Bangkok Metropolitan Region (BMR)[4] faces high flood risk, in particular during the rainy season, between the months of July and October. It experienced major floods in 1942, 1978, 1980, 1983, 1995, 1996, 2002, 2006, and 2011 (Ahsan 2013). Floods have profound short-term impacts on the economy, environment, and public health. They also have long-term economic and social consequences, such as

[4]The BMR consists of the city of Bangkok and five adjacent provinces (Nakhon Pathom, Nonthaburi, Pathum Thani, Samut Prakan, and Samut Sakhon), encompassing 7761.50 km^2.

negatively affecting the attractiveness of expanding existing manufacturing plants or building new ones, and redressing urban poverty and social inequity issues. Because the BMR is projected to undergo rapid demographic, economic, and urban changes, a concerted policy response to the high flood risk it faces can significantly increase its resilience (OECD 2018).

The OECD case study of Bangkok focused on two sets of policy challenges observed in the BMR. First, it analyzed "hard" investments in flood-resilient urban infrastructure integrated with land use planning and zoning policies. Urban infrastructure and land use policies are at the core of flood resilience strategies. Physical capital and the urban form shape the built environment and are major factors exposing land and urban residents to floods. If properly managed, they can, however, be critical in containing such risks, and in contributing to urban green growth (OECD 2018). In the case of Bangkok, the study found that the city of Bangkok and the national government have made massive investments since the 1980s to develop large-scale polder and drainage systems, although they have put much financial burden. To avoid overburdening the city's finances and escalating the need for such infrastructure in the future, the study argues that public authorities need to consider complementing them with a variety of more adaptive infrastructure measures, such as ecosystem-based adaptation or semipermeable surfaces. State-of-the-art information and communication technology (ICT) also offers many opportunities to optimize city functioning, better inform infrastructure decisions, and improve emergency response services, to complement existing flood-control systems (OECD 2018). The study also highlighted the fact that there have been major changes in land use over the past 20 years in the BMR, with significant loss of agricultural lands and conversion to residential, commercial, and industrial uses. Such development has had negative effects on the city's resilience to floods. In particular, it has lost natural water-retention areas (swamps, wetlands, mangroves, etc.) and natural drainage systems that played a key role in managing excess water and limiting the damage done by floods. An efficient region-wide disaster risk management strategy will require coordinating land use strategies and assessing the climate-induced threats, exposure, adaptive capacity, and the resulting vulnerability of urban areas (OECD 2018).

Second, the case study analyzed adoption of "soft" (i.e., nonstructural) resilience measures in the BMR which can not only increase resilience but also help shift its growth model toward a greener, more sustainable development. Engineering solutions to flood risks have been the dominant paradigm for enhancing urban resilience, but in the BMR and elsewhere, soft resilience measures can offer wider opportunities for creating other economic and social benefits and thus help mobilize resources and gain support from other groups in civil society, communities, the media, and the private sector. This will, in turn, avoid economic and/or social resistance to implement policies and programs and eventually reduce the economic and social impact of future disasters. The devastating flood of 2011 in Bangkok demonstrated that the private sector must be targeted and engaged in strategies to increase the city's economic and physical resilience to floods (OECD 2015, 2018). The OECD case study also highlighted the vulnerability of the BMR's poorest residents to extreme weather events. About 73% of people living in low-income communities are affected by the 2011 flood, as they are often located in the most exposed areas near canals and rivers. Local and national governments should find ways to engage local communities in raising public awareness about the need to increase their resilience to floods and other threats. A more consistently integrated and comprehensive approach to vulnerable local communities from governments at all levels would build their social capital, cohesion, and on-the-ground preparedness during a crisis (OECD 2018).

Based on these assessments, the case study recommended following measures for enhancing BMR's urban resilience to floods (OECD 2018):

- Complement existing infrastructure with mangrove forests and wetlands; semi-permeable surfaces on secondary roads and sidewalks with rain-absorbing planter boxes; and urban parks and ecological corridors ("greenways").
- Complement the flood control center with state-of-the-art flood simulation tools, early warning systems, and integrated data collection systems to analyze and disseminate information in real time on changing conditions during times of crisis or urgent need.
- Coordinate flood-resilient building regulations in the entire BMR, and more consistently enforce zoning mechanisms and land use controls to build resilience to floods, such as by more assertively using land value capture tools and financial incentives.
- Develop maps of urban flood risk zones that more accurately reflect changing flooding threats (in terms of exposure and vulnerability) to inform policymakers, residents, and businesses of the potential future impacts.
- Promote insurance mechanisms to build greater financial resilience while supporting green growth policies and programs, business continuity plans, and a database of economic vulnerabilities.
- Use existing community entities, such as schools, religious centers, and public facilities, to increase local communities' resilience to floods and other threats more comprehensively. Engage the media in the coverage of such events and to broadcast public interest messages, community events, and information about flood risks and other impending threats.

9.5 Conclusion

This chapter presented a recent OECD study on disaster risk management in Southeast Asian cities. The study presented a framework for national and subnational governments to assess disaster risk management policies for enhancing resilience in Southeast Asian cities exposed to a variety of natural hazards. Among the key findings of the study, this chapter highlighted policy synergies and complementarities between disaster risk management and urban green growth policies, and recommended integrating disaster risk management into urban green growth policy frameworks. Such an integrated approach can also provide useful insights into the implementation of the Sustainable Development Goals (SDGs), particularly the Goal 11 on "making cities inclusive, safe, resilient and sustainable."

The chapter also summarized the OECD assessment and recommendations of disaster risk management policies in the Bangkok Metropolitan Region (BMR). A more consistent and comprehensive approach was recommended, by combining "hard" investments in flood-resilient urban infrastructure integrated with land use planning and zoning policies and "soft" (i.e., nonstructural) resilience measures, including innovative financing mechanisms and community engagement. The four other case studies (Bandung, Cebu, Hai Phong, and Iskandar), conducted within the same project framework, also provide insightful disaster risk management examples and lessons. Many of these analyzed examples can also be applicable to other regions of the world, although it is important to stress that successful disaster risk management outcomes will require locally tailored models, taking different urban policy contexts into account. In this regard, continuing systematic reviews of urban disaster risk management policies in Asia and other parts of the world can be useful to provide a comparable assessment of the current state of play, consider options to fit for the future, and guide public action and decisions.

References

Ahsan S (2013) Resilient cities for the poor or by the poor? A case study from Bangkok. Technische Universität, Berlin

Daudey L, Matsumoto T (2017) Integrating urban resilience and resource efficiency into local green growth strategies: the case of fast-growing cities in Southeast Asia. Int J Urban Sustain Dev 9(2):226–241. https://doi.org/10.1080/19463138.2017.1339278

IPCC (2012) Managing the risks of extreme events and disasters to advance climate change adaptation. A special report of Working Groups I and II of the Intergovernmental Panel on Climate Change. Cambridge University Press, Cambridge/New York

Lee S (2013) Eco-efficient water infrastructure: towards sustainable urban development in Asia and the Pacific. http://www.unescap.org/sites/default/files/EEWI_Background%20Paper.pdf

Matsumoto T, Daudey L (2014) Urban green growth in dynamic Asia: a conceptual framework. OECD Regional Development Working Papers, No. 2014/12. OECD, Paris, https://doi.org/10.1787/5js7svlw8m0x-en

OECD (2012) Redefining "urban": a new way to measure metropolitan areas. OECD, Paris. https://doi.org/10.1787/9789264174108-en

OECD (2013) Disaster risk financing in APEC economies: practices and challenges. OECD, Paris. http://www.oecd.org/daf/fin/insurance/OECD_APEC_DisasterRiskFinancing.pdf

OECD (2015) Green growth in Bangkok, Thailand, OECD green growth studies. OECD, Paris. https://doi.org/10.1787/9789264237087-en

OECD (2016) Urban green growth in dynamic Asia, OECD green growth studies. OECD, Paris. https://doi.org/10.1787/9789264266360-en

OECD (2018) Building resilient cities: an assessment of disaster risk management policies in Southeast Asia, OECD green growth studies. OECD, Paris. https://doi.org/10.1787/9789264305397-en

OECD Knowledge Sharing Alliance website http://www.oecd.org/knowledge-sharing-alliance/

UN DESA (2014) World urbanisation prospects: the 2014 revision. https://esa.un.org/unpd/wup/Publications/Files/WUP2014-Report.pdf

Chapter 10
Knowledge Co-production Processes for Building Disaster Resilience of Communities in Coastal Areas: A Case Study of Baler, Aurora, Philippines

Pedcris Miralles Orencio

Abstract One of the cornerstones of disaster risk reduction (DRR) is to build the resilience of communities against the negative effects of hazards. Part of this process may entail communities' participation in knowledge co-production and decision making, which can be done through participatory grassroots approaches. To illustrate how strategies for building disaster-resilient communities would benefit from the use of grassroots assessments methods, a real-world application of a participatory approach was undertaken in the five coastal communities of Baler in the province of Aurora in the Philippines. The approach is two-pronged consisting of a survey of communities' perception of disaster resilience, and a participatory workshop for understanding the gaps in the current DRR system. The former was based on a six-level scoring system modified and referenced from the Communities Advancing Resilience Toolkit (CART). The latter, on the other hand, evaluated the four components of disaster-resilient communities using the four resilience dimensions – people, community processes, organizations, and resources. The results revealed that coastal communities were generally not disaster-resilient as reflected by the issues in organizational, financial, infrastructure development, and institutional coordination within the municipality. This attempt to induce a transdisciplinary process has presented useful information sources and provided mechanisms for instilling collaboration among different DRR stakeholders. More importantly, it paved a way for understanding how to use and put value to local knowledge as a source of information in developing strategies for building disaster-resilient communities.

Keywords Disaster resilience · Transdisciplinary · Decision-making process · Grassroots approach

P. M. Orencio (✉)
Southeast Asian Regional Center for Graduate Study and Research in Agriculture (SEARCA), Los Baños, Laguna, Philippines
e-mail: pmo@searca.org

M. Tanaka, K. Baba (eds.), *Resilient Policies in Asian Cities*,
https://doi.org/10.1007/978-981-13-8600-8_10

10.1 Introduction

The substantial damages and setbacks to economic growth and development by disasters have prompted national-level policies to consider the management of associated risks to be integrated with the aspired sustainable development. This has given birth to concepts such as invulnerable development, which looks at the manner to pursue development as to reduce vulnerability (McEntire 2001; Weichselgartner 2002). Blaikie et al. (1994) described vulnerability as a combination of characteristics of a person or group, expressed as an outcome of hazard exposure, which could be derived based on the social and economic susceptibility or condition of the individual, family, or community concerned. Over time, these complex interactions between processes that enhance conditions for hazard, exposure, and vulnerability for society were observed to determine what is known as disaster risks.

The earliest approach that looked at disaster risk from the viewpoint of human context was probably done by White (1945), which eventually led to the development of human participation on disaster risk planning. The UN International Strategy for Disaster Reduction (UNISDR), with its purview on this subject, has put forward in 2004 a systematic development and application of policies, strategies, and practices to minimize vulnerabilities and disaster risks throughout a society, and to avoid or limit the adverse impact of hazards now commonly known as disaster risk reduction (DRR). Among the important cornerstones of DRR is to build the resilience of at-risk communities at the local level (UNISDR 2007). Whereas, a general view on resilience takes into consideration what local communities can do for themselves and how to strengthen their existing capacities, rather than concentrating on people's vulnerability to disaster or their needs in an emergency (Twigg 2007; Olwig 2012).

Despite the foregoing developments, an approach to DRR that can integrate a hazard risk assessment, socioeconomy, and decision-making processes into a more holistic concept has been so far unpopular (Mercer et al. 2008). These are due to some concerns that range from the availability of data for analyzing potential risks at different scales to reliability and accuracy for effective referencing particularly, in developing countries (Orencio et al. 2014). On one hand, the lack of locally based predictions and anticipatory mechanisms has hampered the creation of information that could support a systematic and integrated approach for local adaptation and disaster preparedness (Orencio et al. 2013a). On the other hand, the practice of consensus in tackling a decision in question has always been difficult (e.g., Amason 1996) because strategies tend to adhere with paternalistic and expert-oriented approaches (White et al. 2001). This limits the information and participation to only a few people, which eventually do not guarantee a quality knowledge exchange (Krütli et al. 2006).

Meanwhile, the use of grassroots approach has substantially contributed to complex decision making, as exemplified in some documented experiences of at-risks coastal communities in the Philippines, Japan, and the USA (Orencio et al. 2016). The localized implementation of DRR in these countries has improved because of the promotion of inter-local cooperation as a strategy for responding to certain needs and geophysical peculiarities. As understood, the strategy was feasible owing to several features that were mainly missing from a top-down expert group approach such as tackling a decision in question to go beyond the linear framework of knowledge exchange or classical one-way communication (Krütli et al. 2010). In this process, more people, especially local communities, enjoy the chance to participate in the discussion of risk-related issues (e.g., Krütli et al. 2010) not limited to technological risks (e.g., Renn 2005) through methods that allow for multi-stakeholder participation (e.g., Petts 2004). This led to common knowledge which communities used for determining social protection, resource management, and policy development.

10.1.1 Creating Knowledge from Grassroots Approach for DRR

Knowledge is an intellectual asset for organizations and communities that is found important in a DRR system (Mercer et al. 2008). This includes information, which were shared by all actors in the community, and were readily used in implementing DRR strategies. Ravn (2004) argued that knowledge prospers with interaction and has been influenced by the way actors positioned themselves based on a set of methods of dialogues known as interfaces (Zingerli et al. 2009). The knowledge exchange, as understood, was needed to be two way to allow a joint knowledge construction between diverse actors coming from different practices, which is achieved by positively drawing tacit knowledge from each actor (Polanyi 1983) prior to an explicit all knowledge system (Nonaka 1991; McAdam et al. 2007).

To address the concern on quality knowledge exchanges about disaster planning, DRR experts and practitioners have employed strategies that enhance the sharing of information by engaging multi-stakeholder groups in a participatory process (e.g., Chilvers 2007). This is not so much dependent on the technique applied but rather on the levels of engagements required from participants as well as their attitudes and behaviors towards the discussion at hand (Chambers 1994; Kumar 2002). To increase people participation and knowledge building between expert groups, for instance, Krütli et al. (2006) emphasized transdisciplinary approaches, wherein the interactions between experts and technical people and the grassroots and general public are boosted. This also endeavors academic and nonacademic actor to initiate knowledge co-production (Robinson and Tansey 2006).

Achieving a balance of the produced knowledge was also intricate considering the diversity of sources and forms (Davenport and Prusak 1998). One way is by handling diverse thought collectives and thought styles (Fleck 1979). This process is usually structured and guaranteed by an arena (Pohl et al. 2010), and the produced knowledge is packaged and analyzed for smooth integration to decision making (Bammer 2005; Jahn Hansen 2006). Another way is to use a variation of low- and/

or high-tech tools based on the benefits of applicability to the discussion at hand, and the ability to reduce drawbacks on quality representation and involvement (Coyle 2011). Usually, the idea is to supplement the technical results by engaging a range of stakeholders that could provide further quality and validity to the information production process (e.g., Pretty 1995; Webler 1999).

10.1.2 Knowledge Production Towards a Sustainable DRR Policy

Transdisciplinarity is one of the key features of sustainable development that takes on the process of knowledge co-production (Pohl et al. 2010). Basically, the purpose is to facilitate the knowledge exchange and then determine which information are useful for planning and decision making. Application of transdisciplinary methods (e.g., Klein 2001; Rist and Dahdouh-Guebas 2006; Hardon et al. 2006) places high importance on the interactions between various stakeholders and promotes a considerable degree of flexibility for recognizing respective roles in the deliberation process (Schneider et al. 2009).

For disaster-prone areas like the Philippines, one classic example of a transdisciplinary method is raising public consciousness on the possible impacts of climate-driven sea level rise. The action is linked to the presentation of facts, formulation of timely, responsive, and predictive disaster mitigation plans as well as policy options that allow government to proactively act to potentially-damaged sectors like agriculture and fisheries. In most cases, these policies operate within an overarching framework for enhancing DRR systems at different scales of intervention.

A salient policy that the Philippines has enacted as far as DRR is concerned is the Disaster Risk Reduction Management Act (DRRMA) or Republic Act 10121. This policy aims to strengthen and enhance DRR mechanisms and structures at the sub-national levels. At this scale, the DRRMA outlines the opportunities between institutions for effective transition towards holistic, integrated, and proactive approaches in addressing socioeconomic and environmental impacts of disasters, including those caused by climate variability, by synergizing plans and programs for disaster risk reduction at the national, sectoral, and local levels.

To fast track the implementation of the policy, the Philippine government have suggested collaborations towards continued benchmarking and localized vulnerability studies, and workable institutional arrangements that can achieve a more efficient and cost-effective system for risk information generation (NEDA et al. 2008). Because strategic positioning, Local Government Units (LGUs) are seen as effective instittuions because they can easily identify and appropriately respond to the cause and effects of disasters, and properly consider DRR in their respective local development plans. In terms of implementation, proximity has likewise allowed LGUs to readily provide effective leadership for their citizens, as well as the opportunity to catalyze and sustain behavioral change at individual and community levels, which are necessary elements for mainstreaming community resilience.

This chapter proposed the grassroots approach for knowledge co-production in the development of a viable DRR system that will be in consonance with the Philippines' prevailing local level jurisdictional arrangements (1) by involving multi-stakeholders in assessing local-level DRR initiatives and (2) by formulating strategies for building community resilience that can be used for DRR planning at the local level.

10.2 Methodology

The increasing importance of building resilient communities as a strategy for DRR roots from an understanding that resilient systems are less prone to the negative effects of disasters (Orencio and Fujii 2013b). Bruneau et al. (2003) discussed robustness, redundancy, resourcefulness, and rapidity as key factors to resilient systems. With reference to these factors, Communities Advancing Resilience Toolkit (CART), an intervention to build community resilience to disasters in a form of an assessment, was developed by Pfefferbaum et al. (2011). Meanwhile, Center for Community Enterprise et al. (2000) has considered people, organizations, resources, and community processes as the four dimensions of resilience, wherein the first three are considered the nature of resources while the fourth, community processes, describes the approaches for organizing the other former three.

In this study, a two-pronged approach for knowledge co-production through an assessment method, consisting of a survey and a stakeholder workshop, was conducted to determine the disaster resilience characteristics of coastal communities. The survey was based on a modified version of a scoring system with reference to the CART approach. The workshop, on the other hand, tackled the four dimensions for resilience to understand the emerging and evolving trends for a DRR environment at the local level (Fig. 10.1). The local level, considered in this two-pronged approach, is within the scale that ranges from community level to a town level.

For this purpose, a case study was conducted in the coastal communities of the municipality of Baler, province of Aurora to determine the various actors' level of involvement and empowerment (Arnstein 1969) in the development process and to endeavor pathways for systematizing strategies for multi-stakeholder participation. The assessment for level of knowledge and involvement considered three concepts, namely, (1) information sharing, (2) consultative meeting, and (3) collaborative discussions. These served as bases because of their relative significance to the process of knowledge co-production. The results were then used as inputs in drafting a strategic DRR plan that includes viable modalities for a localized DRR implementation.

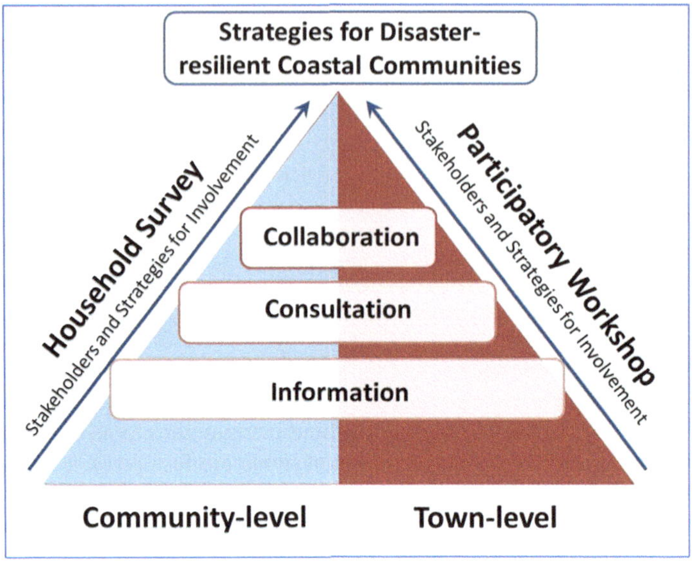

Fig. 10.1 A two-pronged approach for knowledge co-production in developing a strategy for building disaster-resilient coastal communities at the local level

10.2.1 Household-Level Survey

To undertake the survey, this study used a set of desirable criteria for assessing community resilience at the local level based on the study of Orencio and Fujii (2013b). A questionnaire survey in a form of a score-sheet was then developed drawing from a set of 10 process-based and 25 outcome-based indicators (Table 10.1). The outcome-based indicators were represented by four major components, namely, Environmental and Natural Resources Management (ENRM), Sustainable Livelihood (SL), Social Protection (SP), and Planning Regimes (PR). Each of the four components was composed of community-level indicators that were scored based on a six-level scale for ranking activities based on a distinctive risk-reduction intervention. A score of 0 is the lowest as represented by the absence of a clear and coherent activity in an overall DRR program, and score of 5 is the highest as represented by a culture of safety, where DRR is embedded in all relevant policy, planning, practice, attitudes, and behavior (Table 10.2).

In February 2013, a total of 500 score sheets were administered in random at the household level in the five coastal communities of Baler, namely Buhangin, Pingit, Reserva, Sabang, and Zabali, and targeting only the head of the households as respondents. The objective of this activity was to determine the coastal community's resilience character by gathering the households' views and perceptions regarding their inherent capacity for risk reduction as well as the external support they have received for implementing DRR. Based on the survey results, the coastal communities' specific levels of resilience were determined. The scores of the process-

Table 10.1 List of process- and outcome-based indicators used in assessing the level of disaster resilience of coastal communities in Baler, Aurora

Process-based indicators		Outcome-based indicators			
No	Description	Community outcomes		External outcomes	
		No	Description	No	Description
		Environment and natural resources management (ENRM)			
1	Working group for conducting vulnerability and capacity assessments is established	**1**	Understanding of functioning environment and ecosystems	**1**	Supportive policy and institutional structure
2	Baseline survey on community knowledge and capacity has been conducted	**2**	Environmental practices that reduce hazard risk	**2**	Prevention of unsustainable land use
3	A quasi-support group or a community-based action team (CBAT) for disaster response has been formed in the community	**3**	Application of indigenous knowledge and technologies	**3**	Policy linking environmental management and risk reduction
4		**Sustainable livelihood (SL)**			
	The CBAT is mobilized together by a higher level disaster risk reduction (DRR) team	**4**	High level of local economic and employment stability	**4**	Equitable economic development
5	A vulnerability or risk assessment has been conducted in the community	**5**	Livelihood diversification in rural areas	**5**	Diversification of national and subnational economies
6	All people in the community are involved in DRR planning	**6**	Fewer people engaged in unsafe livelihood	**6**	Poverty-reduction targets vulnerable groups
7	Information disseminated prior to DRR planning and risk assessment has been conducted in the community	**7**	Adoption of hazard-resistant agriculture	**7**	Incentives to reduce vulnerable livelihood
8	The community is aware and informed about the DRR activities	**8**	Local market and trade links protected from hazards		
9		**Social protection (SP)**			
	The community is practicing DRR-related planned activities	**9**	Social network systems on DRR activities	**8**	Social safety nets for vulnerable groups
10	The community is involved in the monitoring and evaluation of DRR-related planned activities	**10**	Cooperation with local community for DRR activities	**9**	Comprehensive partnership with external agencies on DRR
		11	Community access to basic social services		

(continued)

Table 10.1 (continued)

		Planning regimes (PR)			
		12	Community decision making takes on land use and hazards	**10**	Compliance with standard international planning
		13	Local community participates in all stages of DRR planning	**11**	Land use planning takes hazard risks into account
				12	Land use plan schemes based on risks assessments

Table 10.2 Scoring system for ranking process- and outcome-based indicators of the four criteria of disaster-resilient coastal communities (based on Orencio and Fujii 2013b)

Score	Description
0	If the respondent is sure that there are *no clear activity or related-activities* conducted by the community or by external institutions
1	If the respondent has mentioned that the community or external institution *respond only during specific crisis situations and with a limited level of awareness and actions*
2	If the respondent has mentioned that community or external institution *plan to address concerns but actions are still short-term or piecemeal and with limited capacity* (e.g., knowledge and skills, material resources)
3	If the respondent has mentioned that *interventions based on plans gear towards long-term development and capacity of the community or external institution to act has improved*
4	If the respondent has mentioned that *interventions are more extensive and these are linked to a bigger coherent long-term strategy that was formulated by community or external institution*
5	If the respondent feels *positive that disaster risk reduction (DRR) has been embedded in all relevant aspects of policy and planning, and is practiced by all stakeholders*

and outcome-based indicators were computed based on steps for measurement by Orencio and Fujii (2013b).

10.2.2 Feedback Session and Stakeholder Workshop

The survey results were analyzed and presented for feedback and validation with the members of the community and local government councils of the five coastal barangays in March 2013. The objective of the activity was to collate people's knowledge, views, and understanding on DRR in order to further elucidate information that will support the objective of localizing the concept. The activity has engaged the participants in a discussion on the current mechanisms for achieving disaster-resilient coastal communities having laid down the assessment of the respective current characteristics of five coastal barangays. This has been subsequently considered as a crucial step for the stakeholder workshop.

The overall expectation from conducting the stakeholder workshop was to create a mental map of a disaster-resilient coastal community. The workshop was conducted in three parts, namely (1) identification of different stakeholders in the local levels and their relevant participation and coordination for implementing DRR, (2) ranking of opportunities and identification of challenges in achieving disaster-resilient communities, and (3) strategizing the four dimensions of resilience (Center for Community Enterprise et al. 2000) with reference to four major criteria for disaster-resilient communities (Orencio and Fujii 2013b). The last part was particularly intended to assess the dimensions and to propose workable mechanisms for an effective strategy for localizing DRR implementation.

10.3 Results

10.3.1 Survey Scores

The scores of the five coastal communities as indicated by the process- and outcome-based indicators in the modified CART questionnaire survey revealed that coastal communities have not achieved significant development in undertaking DRR activities. The indicator measurements showed that all communities fell within the scoring range of zero to one suggesting that there was an absence of clear DRR and DRR-related activities by communities (Fig. 10.2).

Among the five coastal communities, Zabali, Reserva, and Buhangin received the highest computation for the process-based indicators with 0.05, while Sabang and Pingit received the least with 0.04. Both the scores were all nearing zero, which is the lowest score for disaster resilience.

Based on an indicator-specific measure, most coastal communities achieved a low-level development on the indicators for forming and mobilizing a Community-based Disaster Action Team (CBAT) with an average score of 0.09 across the five communities. This was closely followed by the indicator on the conduct of vulnerability or risk assessment and by the indicator for awareness and information on DRR activities with a same score of 0.08. Among the ten process-based indicators observed, the lowest score of 0.00 was received by the indicator on the conduct of information dissemination prior to the DRR planning and by the indicator for conducting a baseline survey of community knowledge and capacity.

In terms of overall outcome-based indicators, Reserva received the highest score with 0.29, while Sabang received the least with 0.21. A score of 0.43, on the other hand, was received by Sabang on the component ENRM, followed by Zabali with 0.40. The barangay that received the least score in this component is Reserva with 0.36. On SL, Zabali ranked first with a score of 0.28, while Buhangin has the least score with 0.22. Reserva, on the other hand, has the highest score 0.35 on the

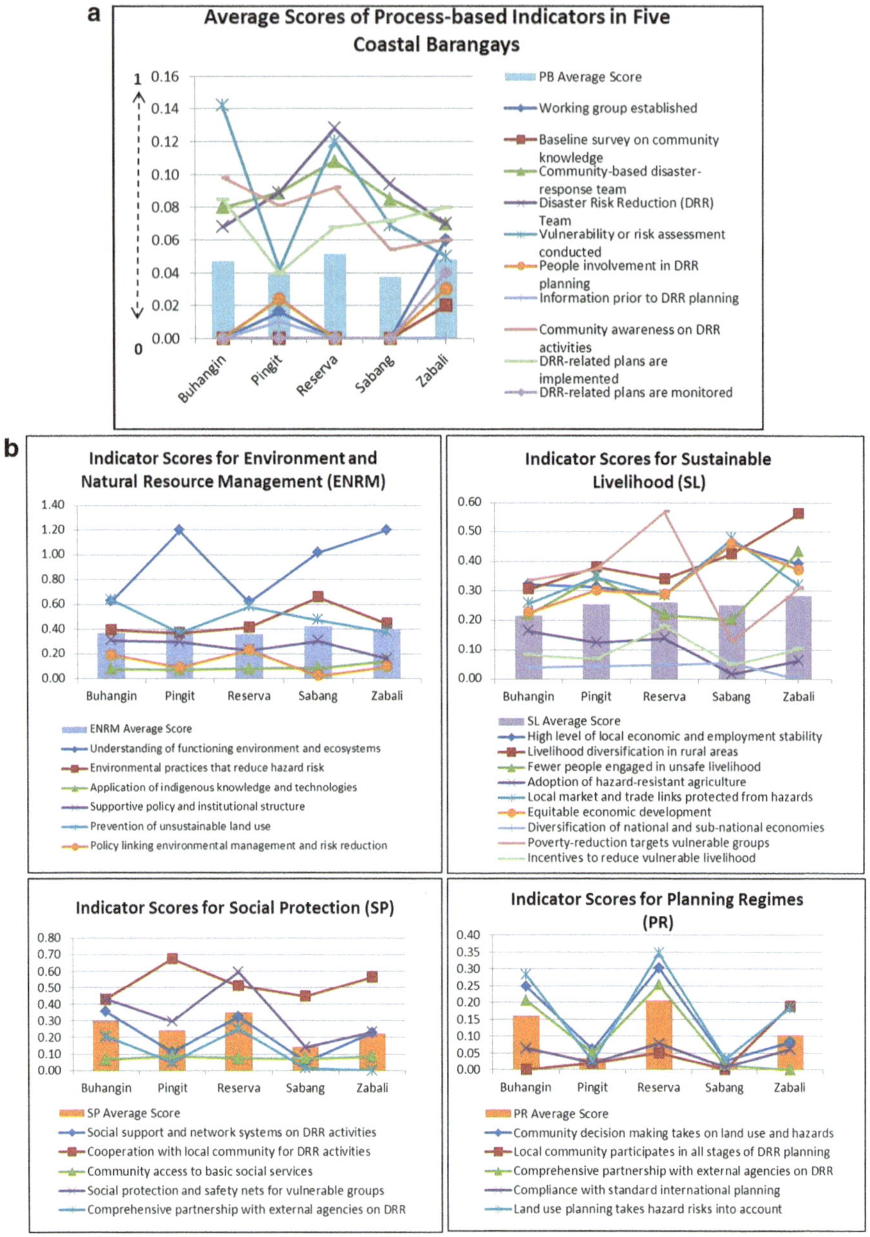

Fig. 10.2 The average scores of process-based indicators (**a**) and outcome-based indicators achieved by community and by externally-led interventions (**b**) based on the collated responses from households range only from 0 to 1, which means that there is an absence of DRR activities in the coastal communities

component SP, while the smallest score of 0.14 was computed for Sabang. A highest score of 0.20 was received by Reserva and the lowest score of 0.10 by Zabali for the PR component.

Among the indicators of the component ENRM, it was the understanding of functional environmental system that received the highest average score of 0.93. The lowest score of 0.09 was given by the respondents to the indicator for the application of indigenous knowledge technologies. For the component SL, the score of 0.34 was the highest and this was computed for the indicator on diversified livelihood activities at the rural level, but the lowest score of 0.04 was given by the respondents to the indicator, diversification of economies in national and subnational levels.

For the SP component, the indicator for cooperation with local community on DRR activities received the highest average score of 0.53, while the lowest score of 0.08 was received by the indicator for community access to basic social services. Finally, for the PR component, the indicator for taking account hazard risk in land use planning received the highest score of 0.17, while the lowest same score of 0.05 was shared by the indicators for local-level participation on DRR planning and compliance with international standards for planning, respectively (Table 10.3).

10.3.2 Workshop Outputs

The participants of the workshop have provided an idea of how stakeholders at the local levels were participating in DRR-related activities in the coastal communities. Using the four major thematic areas in DRR, namely mitigation, preparedness, response, and recovery, the participant jointly determined the different DRR actors. In each thematic area, various offices and departments from the local governments as well as existing councils from communities that were foreseen to be involved in DRR activities were identified (Table 10.4).

The participants have likewise determined how information were gathered, shared, and channeled for undertaking the different stages of DRR (Fig. 10.3). It was found that most information for all aspects of DRR were provided and coordinated by the local DRR office. For instance, in order to undertake preparation for an imminent disaster, the local DRR offices collected vital information from a local weather office. The information included potential weather disturbances that were expected to pass by the coastal areas. Local agricultural and environmental offices, on the other hand, gathered this information for dissemination to sectoral agents, mainly fisheries and agriculture-based organizations, and for use by the local community councils in undertaking the necessary community-based preparations.

During disaster response, all information fed by the sectoral councils to the barangay councils were collated and submitted to the social work department. This department worked closely with the local planning department to provide the immediate services needed by the families such as preparing the schools that normally serve as evacuation centers. Organizations that operated in the community level

Table 10.3 Average scores of process- and outcome-based indicators computed from household level responses per community

Components	Indicators	Average scores per community					Overall average scores
		Buhangin	Pingit	Reserva	Sabang	Zabali	
Process-based (PB)	Working group for conducting vulnerability and capacity assessments is established	0.00	0.02	0.00	0.00	0.06	0.02
	Baseline survey on community knowledge and capacity has been conducted	0.00	0.00	0.00	0.00	0.02	0.00
	A quasi-support group or a Community-based Action Team (CBAT) for disaster response has been formed in the community	0.08	0.09	0.11	0.09	0.07	0.09
	The CBAT is mobilized together by a higher level Disaster Risk Reduction (DRR) Team	0.07	0.09	0.13	0.09	0.07	0.09
	A vulnerability or risk assessment has been conducted in the community	0.14	0.04	0.12	0.07	0.05	0.08
	All people in the community are involved in DRR planning	0.00	0.02	0.00	0.00	0.03	0.01
	Information dissemination prior to DRR planning and risk assessment has been conducted in the community	0.00	0.01	0.00	0.00	0.00	0.00
	The community is aware and informed about the DRR activities	0.10	0.08	0.09	0.05	0.06	0.08
	The community is practicing DRR-related planned activities	0.09	0.04	0.07	0.07	0.08	0.07
	The community is involved in the monitoring and evaluation of DRR-related planned activities	0.00	0.00	0.00	0.00	0.04	0.01
	PB average score	**0.05**	**0.04**	**0.05**	**0.04**	**0.05**	**0.04**
Environment and natural resources management (ENRM)	Understanding of functioning environment and ecosystems	0.63	1.19	0.62	1.02	1.20	0.93
	Environmental practices that reduce hazard risk	0.39	0.37	0.41	0.66	0.44	0.45
	Application of indigenous knowledge and technologies	0.08	0.07	0.08	0.09	0.14	0.09
	Supportive policy and institutional structure	0.31	0.30	0.23	0.30	0.16	0.26
	Prevention of unsustainable land use	0.64	0.37	0.58	0.48	0.38	0.49
	Policy linking environmental management and risk reduction	0.19	0.09	0.23	0.02	0.10	0.13
	ENRM average score	**0.37**	**0.40**	**0.36**	**0.43**	**0.40**	**0.39**

Sustainable livelihood (SL)	High level of local economic and employment stability	0.32	0.31	0.29	0.46	0.39	0.35
	Livelihood diversification in rural areas	0.31	0.38	0.34	0.43	0.56	0.40
	Fewer people engaged in unsafe livelihood	0.22	0.35	0.22	0.20	0.43	0.29
	Adoption of hazard-resistant agriculture	0.16	0.12	0.14	0.02	0.06	0.10
	Local market and trade links protected from hazards	0.26	0.35	0.29	0.48	0.32	0.34
	Equitable economic development	0.23	0.30	0.29	0.46	0.37	0.33
	Diversification of national and subnational economies	0.04	0.05	0.05	0.06	0.00	0.04
	Poverty-reduction targets vulnerable groups	0.34	0.37	0.57	0.13	0.31	0.34
	Incentives to reduce vulnerable livelihood	0.08	0.07	0.18	0.05	0.10	0.10
	SL average score	**0.22**	**0.26**	**0.26**	**0.25**	**0.28**	**0.25**
Social protection (SP)	Social support and network systems on DRR activities	0.36	0.11	0.32	0.05	0.23	0.21
	Cooperation with local community for DRR activities	0.43	0.68	0.51	0.45	0.56	0.53
	Community access to basic social services	0.07	0.09	0.08	0.07	0.09	0.08
	Social protection and safety nets for vulnerable groups	0.43	0.30	0.59	0.14	0.23	0.34
	Comprehensive partnership with external agencies on DRR	0.21	0.05	0.25	0.01	0.00	0.10
	SP average score	**0.30**	**0.24**	**0.35**	**0.15**	**0.22**	**0.25**
Planning regime (PR)	Community decision making takes on land use and hazards	0.25	0.06	0.30	0.03	0.08	0.14
	Local community participates in all stages of DRR planning	0.00	0.02	0.05	0.00	0.19	0.05
	Comprehensive partnership with external agencies on DRR	0.21	0.05	0.25	0.01	0.00	0.10
	Compliance with standard international planning	0.06	0.02	0.08	0.01	0.06	0.05
	Land use planning takes hazard risks into account	0.28	0.03	0.35	0.03	0.18	0.17
	PR average score	**0.16**	**0.04**	**0.21**	**0.02**	**0.10**	**0.10**
Resilience	**Overall average score**	**0.26**	**0.23**	**0.29**	**0.21**	**0.25**	**0.25**

Table 10.4 Stakeholders identified under the four thematic areas of DRR in coastal communities

Mitigation	Preparedness	Response	Recovery
Local DRR Office Weather Bureau Local Media	Local DRR Office Local Media Agriculture Office Environmental Office Barangay Council Fisheries Council Agriculture Council Disaster-affected Families	Local DRR Office Civil Defense Office Police Department Budget Department Disaster-affected Families Community Media NGO/NPO Schools Barangay Health Office People's Organization Philippine Red Cross	Local DRR Office Armed Forces Group Civil Society Group Social Work Department Public Works Office Local Planning Department

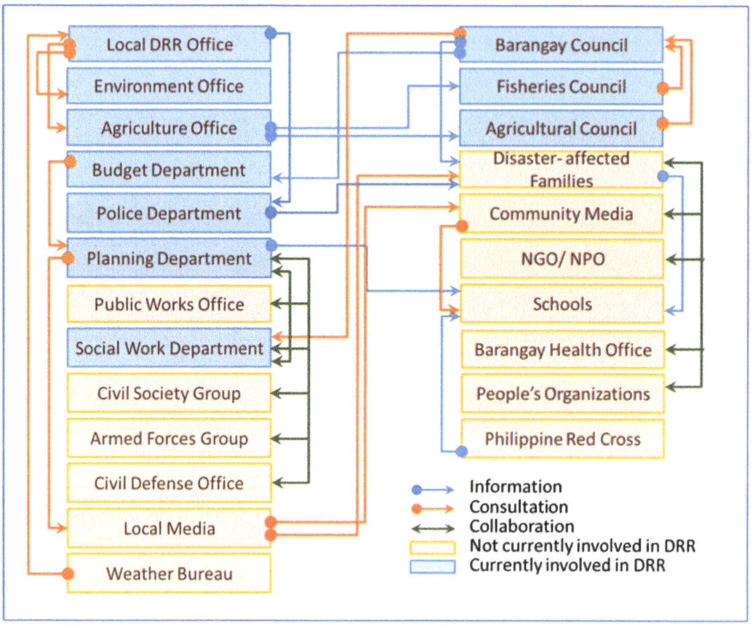

Fig. 10.3 The local stakeholders involved in undertaking DRR in the coastal communities of Baler, Aurora and the relative processes involved when exchanging or sharing DRR-related information

such as nongovernment organizations (NGOs), people's organizations, the health department, and community media groups worked together to guide and support families on undertaking their response activities. During this time, the local DRR office usually tapped the police department in ensuring the safety of livelihood and properties of affected families. Meanwhile, other NGOs like Red Cross also provided relief services. During recovery, institutions at the local level, consist mainly by local planning and social work departments, orchestrated the joint actions of civil society groups, armed forces, and public works in restoring the social and public infrastructures damaged by disasters.

Additional details on the level of disaster resilience of coastal communities also surfaced during the assessment of opportunities and limitations in implementing local-level DRR. During the plenary, the participants altogether ranked the opportunities and limitations and identified fund allocation for DRR and people's low level of knowledge on DRR as the most important concerns (Table 10.5). Finally, during the last segment of the workshop, the participants assessed the four important components for a disaster-resilient coastal community, ENRM, SL, SP, and PR, by looking at the four dimensions of resilience namely, people, process, organization, and resources, in each component. The results of the assessment can be observed in Table 10.5.

10.4 Discussion of Results

10.4.1 Participation through Information and Communication

The use of the two-pronged method in generating information for building disaster resilience has illustrated ways to understand how knowledge production processes can be conducted in the coastal communities. As observed, different stakeholders in a community have different views, experiences, and perceptions. When collated, this can be potent sources of valuable information, especially for planning and strategizing DRR interventions (Mercer et al. 2008). The quality of the information, however, depends on which types of approaches were used. For instance, in this exercise, the use of social survey as a tool for understanding the level of disaster resilience of coastal communities was practically advantageous because of the previous activities on building the capacities of the communities on disaster events. It provided the reference on where and how data should be collected. In this sense, the summary of household level scores that resulted from the survey can easily be treated and accepted as a community condition. Not to mention that such a case of vague or questionable information from a community was even validated during the stakeholder workshop.

The information generated from the survey have been useful reference points for the discussions during the stakeholder workshops. Although the information is not comprehensive because of limitations in terms of the scope and coverage, it has provided the evidences and has showcased proofs that communities can easily relate to when they begin sharing about their current concerns and conditions. This justifies the need for having that information available prior to any community decision-making activity. The real knowledge exchange, however, was evident during the stakeholder workshop. People coming from different background, experiences, professions and mandates look at this as an opportunity to discuss and to agree and

Table 10.5 The disaster resilience of coastal communities in Baler, Aurora based on the four components of disaster-resilient coastal communities and the four dimensions of resilience

Components of disaster-resilient coastal communities	Four dimension of resilience			
	People	Process	Organizations	Resources
Environment and Natural Resources (ENRM)	Local communities; local government officials; students	Implementation of localized environmental laws (ecological solid waste management; comprehensive land use plan; forest land use plan; environmental code; clean air act; coastal resource management plan)	Fisheries and Aquatic Resources Management Council (FARMC); Department of Environment and Natural Resources (DENR); Environment and Natural Resources Office (ENRO); nongovernment organizations (NGOs); academe; environmental law enforcement; small and medium enterprises (SME)	Funds for operations; environmental staff; waste collection facilities and equipment (i.e., patrol boat and vehicle)
Sustainable livelihood (SL)	Community members (fisherfolk; farmers; women; fish and vegetable vendors)	Organize beneficiaries; assess and plan the viability of projects; propose and prepare design of livelihood projects for fund sourcing; implement, monitor, and evaluate the implementation of the project; train beneficiaries on financial management; conduct of impact assessment	Department of Social Work and Development (DSWD); Department of Agriculture (DA); Bureau of Fisheries and Aquatic Resources (BFAR); Department of Trade and Industry (DTI); Department of Science and Technology (DOST); Department of Labor and Employment (DOLE)	Guidance from private sector providers; capacity building of local communities; fund for operations

Social protection (SP)	Local community officials; local government officials; NGOs; police	Provide intelligence reports; installation of early warning systems (EWS); identification of hotspots; regular patrolling; strict enforcement of the environmental and social laws (i.e., ESWM; riverbank easement; hazard zone areas)	Tricycle Operators and Drivers Association (TODA); NGO; media (i.e., radio groups); religious organizations (i.e., church; Knights of Columbus)	Funds for operations; people support to operations; operational materials (i.e., medical supplies; communication and transportation equipment)
Planning regimes (PR)	Local community officials; NGO representatives; Municipal Development Council members (MDC); planning and development officers (i.e., municipal and provincial); budget officer; DRRM officer	Data gathering and assessment; creation of technical working group (TWG) or ad hoc committees; regular monitoring and evaluation, planning and legitimization of programs and activities; advocate resources sharing with other government offices	DENR; external NGOs; foundations; DOST; Red Cross; DSWD; Department of Public Works and Highways (DPWH); Department of Health (DOH); Department of Energy (DOE)	Funds for operations; private service provider; planning systems; coordination mechanisms and equipment; staffing

disagree on various issues, validate the information available, and identify ways to arrive at a resolution. Providing such a venue for discussion does not only compel local leaders to address community issues, it likewise provide the opportunity for community members to participate and be involved in the decision-making process. The results of the workshop, for instance, noted some stakeholders as significant actors in the implementation of DRR in the coastal areas but were not yet part of the existing DRR system. In the current system, information and stakeholder participation are still one-way. Not only that the activities were limited to preparedness, but there is the lack of quality interaction and collaboration among institutions. As described by the flow of information (Fig. 10.3), the major stakeholders in this DRR process were local offices at the municipal level that channeled the information to the community councils which at the community level when received, household-level preparatory activities were undertaken. As observed, the processes were basically a give-and-receive action between external organizations that provide the information and communities that receive and use them for activities.

As evidently shared by the workshop participants, the local planning office can support the local DR office in its role as the focal organization for orchestrating the community actions for implementing local DRR activities, particularly on the disaster recovery process. On the other hand, preparing for imminent disasters is determined to be more community-led especially during pre-emptive evacuations. Barangay health offices through its health workers can lead in the process of evacuation with support from nongovernment organizations (NGOs). Although a systematic response and recovery were not yet in place, potential collaboration can be expected from these actions.

One aspect of the DRR that was also identified absent in the coastal communities is having a program for mitigation. Mitigation can be structural, which involves infrastructure development that complies with planning standards and ensures communities will be protected from the harmful effects of hazards. This can also take the form of natural buffering by protecting and strengthening the coastal areas' ecological systems such as the forest, riverine systems, mangrove areas, wetlands, and dune and coral reef systems. The coastal communities should capitalize on the community-based processes identified under the component ENRM since managing the environment can cut across all levels of governments (Table 10.5).

10.4.2 Stumbling Blocks in the Knowledge Production Processes

Unlike information and consultation, collaboration encourages a higher form of participation between and among stakeholders and therefore harbors more valuable knowledge. However, the results of the analysis of the capacities of coastal communities showed that communities still lacked the required knowledge and intervention when assessed against the qualities of resilient communities. For instance, community-based processes that were indicative of a successfully performing disaster-resilient coastal community were observed to be inefficient. The absence of mechanisms for engaging organizations and institutions in undertaking DRR makes DRR activities difficult to orchestrate. All these situations defeat the objective of enhancing the knowledge sharing between stakeholders.

Generally, the above limitations pose a challenge to substantially identifying the strategies for building disaster-resilient coastal communities. To avoid future shortcomings, it is hereby suggested to enhance the process by (1) improving the key questions in the survey; (2) guaranteeing the representativeness of the grassroots participants in the workshop; and finally, (3) executing the steps towards generating consensus on the strategies considering the background of various sectors as warranted by a participatory approach.

As observed, coastal communities were not at all weak in a sense because they have shown stability and strength on some aspects that DRR actors and local governments could consider in developing them into disaster-resilient communities. There were, however, concerns on attributing the usual community activities as DRR-related. This is evident from the score of zero that indicated that there most respondents don't recognize the DRR activities. Some communites that provided an above zero score to DRR-related initiatives had been involved by different organizations, academe, and national government with communities that preceded this research.

Some communities tend to relate more to the functioning environment and ecosystems. Even the assessment of the four dimensions of resilience for the component ENRM revealed that there were already plans and programs being implemented locally and these were based on national-level environmental laws (Table 10.5). This has been observed to be always in their priorities to which communities can easily take part with. Local DRR actors can turn this into an opportunity for enhancing DRR actions on the ground by developing pragmatic approaches for mainstreaming DRR into the ENRM activities. But just like any other components, sustaining the activities is always affected by the lack of funds, poor staffing, and insufficient equipment. More to improving the conduct of the survey, however, is enhancing the stakeholder workshop by ensuring the quality participation of relevant stakeholders. This starts with the process of identifying the right stakeholder by the convener or initiator to maximize the opportunity for quality knowledge exchange. Understanding which issue or concern will be deliberated is important in identifying the participants, to which a prior consultation with relevant institutions may be conducted. Then during the workshop, to get everyone on board the process, relevant information should be shared. Other key factors that should be taken into consideration include the balance representation from various groups, a neutral ground as the venue, and the discussion time allotted for arriving at a consensus. Allowing the stakeholder to understand and approve the process before conducting the activity is also one way to get them more involved. Finally, for a more objective process, the results of the discussions are to be documented and approved by the participants.

10.4.3 Developing the Strategies for Building Disaster Resilience

The results of the household-level responses suggest that coastal communities were generally not disaster resilient (Score of 0 to <1). No significant disparities on the scores, which range from 0.21 to 0.29, can be observed. In general, this scoring pattern means that the coastal community did not observe nor implement activities for DRR. Neither there were DRR interventions that the communities could participate into because of the prevailing low level of community engagement. While an approach for implementing DRR is hereby suggested to be enforced on all process- and outcome- based indicators in order to build a disaster-resilient community, some critical areas that should be explored when building coastal community resilience must be considered, as these were found to be lacking in all coastal communities.

For process-based indicators, the priority activities should center on (1) the conduct of baseline information on community knowledge and capacity, and (2) the conduct of information dissemination prior to DRR planning and risk assessment. The two activities were found related as they both require working with the communities for collection and distribution of vital information in order to facilitate monitoring and planning. Moreover, both activities consider the gathering of related information that can be used for an assessment. The baseline provides information that were usually used as basis for constructing a picture of the current DRR intervention in the coastal communities, while the type of information for encouraging participation can be used to determine the level of community participation on DRR.

In terms of outcome-based indicators, which were taken from the four components of disaster-resilient coastal communities in Baler, Aurora (Orencio and Fujii 2013b), there were five major areas that should be prioritized. First, there should be diversification of national and subnational economies (Component: SL). Diversification comes in a number of ways – from the expansion of local economic production to an increase in number of income sources to augment the common economic activities in the coastal areas. It has been understood that a lack in diversified economies will contribute to unsustainable practices. For instance, this will encourage competition and encroachment of territorial boundaries, which results when members of the communities, in dire need to support their family needs, struggle to ensure access and utilization to resources, especially, when major economic activities such as fisheries and agricultural industries become affected by natural and man-made changes.

Second, there should be participation of communities in all stages of DRR planning (Component: PR). In general, ensuring participation from different members of the communities is difficult. From the perspective of a local community member, this difficulty can be due of the lack of individual resources to participate – time, interest, and information. In most cases, it is difficult to engage them because community-level endeavor competes with the household-level activities, especially when the latter pertains to economic and livelihood needs. Persuading communities into attending or participating in community activities should nonetheless be backed by quality information. The effectiveness of information will depend on how this will be delivered to the coastal communities.

Information is also tied to the third concern, which is compliance to international planning standards (Component: PR). Among the final steps in building coastal community resilience is to strengthen the priorities that were identified in the process of decision making through planning and programming. Planning is a way to bring people together and think of different options that communities would like to implement. A plan therefore must be customized based on the needs and specific priorities of the people comprising the communities. The process of strategizing and priority setting, however, must always be guided by standards that provide international benchmarks and targets because this will help communities to uniformly assess their conditions and progress against other deliverables over time.

The fourth area for consideration in building coastal community resilience is community access to basic social services (Component: SP). The proliferation of limited access to basic services and productive assets has been considered as persistent challenges for development in the coastal communities. This condition was more prevalent in poor and underserved households because of the substantial fragility of living conditions. For instance, these households were generally the ones with no electricity source, unsafe water sources, or have lesser capacity to attend school. The situation has been exacerbated by limited opportunities to achieve a generally secured employment, a relatively viable livelihood options and adequate services, and support from governmental institutions. The objective therefore is to strengthen community-level partnerships and empower communities by enhancing delivery of service and development opportunities that reduce poverty.

Finally, the fifth area for enhancing coastal community resilience is to apply Indigenous knowledge and technologies (Component: ENRM). Indigenous knowledge and technologies essentially forms the cultural practices, traits, and behavior that is peculiar to a certain community. In terms of the coastal communities under study, community members have adaptation mechanisms that were not particularly observable in other areas. This tacit knowledge is difficult to codify and may require an in-depth assessment as to how the communities come up with ways to interact and react with the changes in their environment. Because indigenous knowledge is also adoptive and specific to the conditions in their communities, people could respond uniquely to the changing economic, social, ecological circumstances. Tapping into what is local and inherent could therefore be useful in implementing a localized decision-making approach for building a resilient coastal community.

10.5 Conclusion

The two-pronged method has presented a systematic participatory approach for collecting and managing the perception and experiences of coastal communities, and these were found to be good inputs for planning a localized DRR system. First, by referring to knowledge exchange process and the knowledge produced from the local communities, the major stumbling blocks for implementing DRR at the local level can be determined. Second, the focus and the kind of information that were targeted by the approach could also prompt potential collaboration among other stakeholders, which are important when undertaking further risk assessments. Third, a collaborative scheme that encourages participation from other organizations will bring in a systematic generation of local knowledge that can be used for managing potential risks of coastal communities to disaster-causing hazards.

References

Amason AC (1996) Distinguishing the effects of functional and dysfunctional conflict on strategic decision making: resolving a paradox for top management teams. Acad Manag J 39(1):123–148
Arnstein SR (1969) A ladder of citizen participation. J Am Inst Plann 35(4):216–224
Bammer G (2005) Integration and implementation sciences: building a new specialization. Ecol Soc 10(2):6
Blaikie P, Cannon T, Davis I, Wisner B (1994) At risk: natural hazards, people's vulnerability, and disasters, 1st edn. Routledge, London/New York
Bruneau M, Chang SE, Eguchi RT, Lee GC, O'Rourke TD, Reinhorn AM, von Winterfeldt D (2003) A framework to quantitatively assess and enhance the seismic resilience of communities. Earthquake Spectra 19(4):733–752
Chambers R (1994) The origins and practice of participatory rural appraisal. World Dev 22(7):953–969
Chilvers J (2007) Towards analytic deliberative forms of risk governance in the UK? Reflecting on learning in radioactive waste. J Risk Res 10(2):197–222
Centre for Community Enterprise, Lewis M, Rowcliffe P (2000) The community resilience manual: a resource for rural recovery and renewal. Centre for Community Enterprise, Port Alberni
Coyle SJ (2011) Sustainable and resilient communities: a comprehensive action plan for towns, cities, and regions, vol 15. Wiley, Hoboken
Davenport TH, Prusak L (1998) Working knowledge: how organizations manage what they know. Harvard Business Press, Boston
Fleck L (1979) Genesis and development of a scientific fact. University of Chicago Press, Chicago (First published in German, 1935)
Hardon GH, Bradley D, Pohl C, Rist S, Wiesmann U (2006) Implications of transdisciplinarity for sustainable research. Ecol Econ 60(1):119–128
Jahn Hansen C (2006) Urban transport, the environment and deliberative governance: the role of interdependence and trust. J Environ Policy Plan 8(2):159–179
Klein JT (2001) The discourse of transdisciplinarity: an expanding global field. In: Transdisciplinarity: joint problem solving among science, technology, and society. Birkhäuser, Basel, pp 35–44
Krütli P, Stauffacher M, Flueler TM, Scholz RW (2006) Public involvement in repository site selection for nuclear waste: towards a more dynamic view in the decision-making process. In: Conference proceedings. VALDOR 2006—VALues in decisions on risk. Stockholm, May 14–18, 2006. SKI, SEPA, SGI, SRCE, OECD/NEA, UK Nirex, pp 96–105
Krütli P, Stauffacher M, Flüeler T, Scholz RW (2010) Functional – dynamic public participation in technological decision – making: site selection processes of nuclear waste repositories. J Risk Res 13(7):861–875
Kumar S (2002) Methods for community participation: a complete guide for practitioners. ITDG, London

Mercer J, Kelman I, Lloyd K, Suchet-Pearson S (2008) Reflections on use of participatory research for disaster risk reduction. Area 40(2):172–183
McAdam R, Mason B, McCrory J (2007) Exploring the dichotomies within the tacit knowledge literature: towards a process of tacit knowing in organizations. J Knowl Manag 11(2):43–59
McEntire DA (2001) Sustainability or invulnerable development? Proposals for the current shift in paradigms. Aust J Emerg Manag 15:1 58–1 61
Nonaka I (1991) The knowledge-creating company. Harv Bus Rev 69(6):96–104
National Economic and Development Authority (NEDA), UNDP and ECHO (2008) Mainstreaming disaster risk reduction in sub-national development and land use/physical planning in the Philippines, ISBN 978-971-8535-23-3
Olwig M (2012) Multi-sited resilience: the mutual construction of "local" and "global" understandings and practices of adaptation and innovation. Appl Geogr 33:112–118. https://doi.org/10.1016/j.bbr.2011.03.031
Orencio PM, Fujii M (2013a) An index to determine vulnerability of communities in a coastal zone: a case study of Baler, Aurora, Philippines. AMBIO J Hum Environ 42(1):61–71
Orencio PM, Fujii M (2013b) A localized disaster-resilience index to assess coastal communities based on an analytic hierarchy process (AHP). Int J Disaster Risk Reduct 3:62–75
Orencio PM, Fujii M (2014) A spatiotemporal approach for determining disaster-risk potential based on damage consequences of multiple hazard events. J Risk Res 17(7):815–836
Orencio PM, Endo A, Taniguchi M, Fujii M (2016) Using thresholds of severity to threats to and the resilience of human Systems in Measuring Human Security. Soc Indic Res 129(3):979–999
Republic Act No. 10121. An act strengthening the Philippine disaster risk reduction and management system, providing for the national disaster risk reduction and management framework and institutionalizing the national disaster risk reduction and management plan. Accessed on 29 May 2013 from http://www.ndrrmc.gov.ph/index.php?option=com_content&view=article&id=45:republic-act-no-1
Ravn JE (2004) Cross-system knowledge chains: the team dynamics of knowledge development. Syst Pract Action Res 17(3):161–175
Renn O (2005) Risk governance towards an integrative approach. White paper no. 1, Sep. 2005. IRGC. Geneva. Accessed on 27 Aug 2015 from http://www.irgc.org/IMG/pdf/IRGC_WP No 1_Risk_Governance__reprinted_version_pdf
Rist S, Dahdouh-Guebas F (2006) Ethnosciences – a step towards the integration of scientific and indigenous forms of knowledge in the management of natural resources for the future. Environ Dev Sustain 8(4):467–493
Pfefferbaum RL, Pfefferbaum B, Van Horn RL (2011) Communities advancing resilience tool kit (CART): the CART integrated system. Terrorism and Disaster Center at the University of Oklahoma Health Sciences Center, Oklahoma City
Pohl C, Rist S, Zimmermann A, Fry P, Gurung GS, Schneider F, Wiesmann U (2010) Researchers' roles in knowledge co production: experience from sustainability research in Kenya, Switzerland, Bolivia and Nepal. Sci Public Policy 37(4):267–281
Polanyi M (1983) The tacit dimension. Peter Smith, Gloucester
Petts J (2004) Barriers to participation and deliberation in risk decisions: evidence from waste management. J Risk Res 7(2):115–133
Pretty JN (1995) Participatory learning for sustainable agriculture. World Dev 23(8):1247–1263
Robinson J, Tansey J (2006) Co-production, emergent properties and strong interactive social research: the Georgia Basin futures project. Sci Public Policy 33(2):151–160
Schneider F, Fry P, Ledermann T, Rist S (2009) Social learning processes in Swiss soil protection – the 'from farmer-to farmer' project. Hum Ecol 37(4):475–489
Twigg J (2007) Characteristics of a disaster-resilient community: a guidance note, Version 1. UK Department for International Development's Disaster Risk Reduction Interagency Coordination Group, London. Accessed on 22 Nov 2012 from https://practicalaction.org/docs/ia1/community-characteristics-en-lowres.pdf

UN International Strategy for Disaster Reduction (UNISDR) (2004) Living with risk: a global review of disaster reduction activities, 2004. 429 p. ISBN/ISSN: 9211010640
UN International Strategy for Disaster Reduction (UNISDR) (2007) Building disaster resilient communities good practices and lessons learned, a publication of the "global network of NGOs" for disaster risk reduction Geneva, 67 p. Accessed on 22 Nov 2012 from http://www.unisdr.org/we/inform/publications/596
Webler T (1999) The craft and theory of public participation: a dialectical process. J Risk Res 2(1):55–71
Weichselgartner J (2002) About the capacity to be wounded: the need to link disaster mitigation and sustainable development. Extreme Naturereignisse–Folgen, Vorsorge, Werkzeuge, DKKV, Bonn, pp 150–158
White GF (1945) Human adjustment to floods: a geographical approach to the flood problem in the United States. Research Paper Number 29 Department of Geography, University of Chicago.
White GF, Kates RW, Burton I (2001) Knowing better and losing even more: the use of knowledge in hazards management. Global Environ Change B Environ Hazard 3:81–92
Zingerli C, Michel C, Salmi A (2009) On producing and sharing knowledge across boundaries: experiences from the interfaces of an international development research network. Knowl Manag Dev J 5(2):185–196

Part III
Recommendations

Chapter 11
"Covenant of Mayors Japan": Regional Re-creation and Global Contributions Through Climate Policy and Energy Autonomy

Noriko Sugiyama

Abstract The Covenant of Mayors (CoM) initiative has been conducted by the EU from 2008, developing into the Global Covenant of Mayors (GCoM), the largest global coalition of cities and local governments voluntarily committed to actively combating climate change. CoM Japan was established by Nagoya University in 2015 and has acted as the regional chapter of GCoM since 2017. Challenges are expected for non-state actors, such as local governments, in the transition to low/de-carbon and climate-resilient cities.

Keywords Climate change · Energy autonomy · Covenant of Mayors · Local climate policy · Japan

11.1 Background

Based on the Act of Promotion of Measures to Prevent Global Warming, established in 1998, local Japanese governments are obligated to formulate regional implementation plans to counter global warming for any core city or large municipality. As of May 2017, 387 local governments had established such implementation plans (Regional Policy Series; Ministry of Environment 2017). However, even with repeated formulation and revision of implementation plans, almost no local governments have been able to realize drastic reductions in greenhouse gas emissions. The primary climate policy measures implemented by local Japanese governments have leaned toward public awareness and information provision; energy, transportation, structure, and community development measures resulting in tangible reductions in CO_2

N. Sugiyama (✉)
Education and Research Center for Sustainable Co-Development, Graduate School of Environmental Studies, Nagoya University, Nagoya, Aichi, Japan
e-mail: n.sugiyama@nagoya-u.jp

M. Tanaka, K. Baba (eds.), *Resilient Policies in Asian Cities*,
https://doi.org/10.1007/978-981-13-8600-8_11

emissions have been few and far between (Sugiyama and Takeuchi 2008). This is largely due to local governments in Japan not having jurisdiction over energy policy.

Moreover, it has been recently estimated that 896 municipalities – especially provincial cities and villages – will disappear as a result of exhaustion of regional economies, rapid aging of the population, and depopulation (Masuda 2014). Faced with a variety of challenges, local governments might be able to create and implement regionally resilient regional development by establishing regional climate measures and energy autonomy. In this section, we first introduce the "Covenant of Mayors", a European mechanism for propagating climate and energy policy initiatives created by local governments. We then describe Japan's "Covenant of Mayors (Mayors of municipalities)", which was created based on the European model.

11.2 Europe's Covenant of Mayors

The "Convent of Mayors" (hereinafter referred to as "CoM") was launched in the European Union (EU) in 2008 from an initiative of the European Commission, and as of June 2017, over 7000 local governments were independently participating within this framework. The heads of participating local governments pledged to strive to reduce CO_2 emissions beyond the levels published by the EU. Local governments that entered into this covenant drafted "Sustainable Energy Action Plans" (hereinafter referred to as SEAP) to help them achieve their goals, implement concrete measures, and submit monitoring reports every 2 years. The CoM was a voluntary initiative with absolutely no obligation to participate, nor will any penalties be imposed should objectives not be met.

When launched in 2008, one goal was to exceed the EU's reduction objective of 20% by 2020, submitting an SEAP within 1 year after entering into the covenant and implementation reports every 2 years. The goals were modified in 2015 to exceed the new EU target of a 40% reduction in greenhouse gas emissions by 2030, and added adaption measures for climate change and access to safe and affordable sustainable energy. This format integrated a Covenant of Mayors concerning adaptation measures for climate change (Mayors Adapt), initiated in 2014. Local governments with covenants were to establish a Sustainable Energy & Climate Action Plan (hereinafter referred to as SECAP) within 2 years of entering the covenant.

Regional, state, and prefecture governments can register as coordinators, offering data to local governments in the region, providing staff member training, and supporting participation in CoMs. Furthermore, international NGOs and local government networks can register as supporters, provide advice and tools to local governments that are considering a CoM, and otherwise fulfill the helpdesk role. Presently, registered CoM coordinators and supporters have increased to 204 and 106, respectively.[1]

In addition, various subsidy payments have been provided for CoMs at the EU level, and local governments that have entered into covenants can receive CoM

[1] Covenant of Mayors website: http://www.covenantofmayors.eu accessed July 2017.

coordinator and supporter aid and are able to apply for assistance payments. The CoM's support systems have resulted in encouraging the participation of many small local governments which have been unable to implement their climate policy against their willingness because the support systems is considered to make up for their lack of information, human resources and budget.

For example, in Chieti Province in the Abruzzo region of Italy (population: 400,000), ALESA – a special organization dedicated to energy and the environment – was created[2] and registered as a CoM coordinator in 2009. ALESA requested 104 local governments in the province to participate in CoMs; 3 months later, all local governments in Chieti had signed a CoM. Thereafter, ALESA met with local government supervisors individually to advance and promote SEAP. Energy conservation improvements to schools and other public facilities in Chieti, including replacing light bulbs and implementing solar [power] systems that provide electricity and heat, were funded from such EU financial mechanisms as ELENA (European Local Energy Assistance) and ERDF (European Regional Development Fund). Consequently, CO_2 emissions from local governments, i.e., the public sector, were cut by about 33%. For local governments, the most attractive aspect was that they could achieve broad reductions in energy expenditures. With these reductions, local governments were able to use the remaining funds for social welfare and other areas.

There have been similar cases where the prefecture promoted local government participation as CoM coordinators in the regions of Turin, Italy and Barcelona, Spain.[3]

The CoM website publishes information from local governments in the covenant, allowing visitors to browse an overview of local governments, CO_2 emissions, other metadata, and major SEAP/SECAP policies. Moreover, outstanding local government initiatives that include joint ownership of content can be searched by area. The website acts as a platform for exchanging information and experiences.

Following the establishment of CoMs, there has been a rapid increase in the number of local governments that have greatly surpassed initial official expectations, i.e., such municipalities exceeded 7000 in 2016. While CoM reports show that this framework involved voluntary participation, it was supported by the strong political commitment from mayors. This facilitated replacing covenants with tangible policies to advance climate change mitigation and adaption measures, which was acknowledged as a model for multilevel governance (Joint Research Centre 2016). In actuality, according to 315 monitoring reports submitted from local government in covenants, greenhouse gas emissions decreased 23% on average. Areas that were most effective in reducing CO_2 emissions were those related to heat supply, such as reductions in energy expenditures as a result of improvements in cooling and heating efficiency for buildings, heat supply in district heating and cooling (DHC), and the introduction of regional heat supply for distribution from

[2]Agenzia Locale Per L'Energia e lo Sviluppo Ambientale; in English "Local Agency for Energy and Environment."

[3]For more details, see Takashi Tanaka's compilation "Implementing Low Carbon/Energy Policy Beginning at the Regional Level" (2014) administration, Chap. 12 "CoM Initiatives in Europe: Formulating Sustainable Regional Energy Plans" (pp. 253–280).

renewable energy and similar measures (36%). The area with the next largest degree of effectiveness was energy conversion (17%). In summary, SEAP/SECAP resulted in policies and ventures aimed at specifically creating a low-carbon society, resulting in a decrease in CO_2 emissions.

11.3 Proposal for a Japanese "Covenant of Mayors"

Based on research on regional climate policy implemented by local governments in Japan, Nagoya University investigated Europe's CoM efforts, and proposed a Japanese "Covenant of Mayors (Mayors of cities/towns/villages)" (hereinafter referred to as "CoM-Japan") in January 2015. The covenant proposed for the CoM-Japan can be classified as (1) local energy production for local consumption, (2) drastic reductions in greenhouse gases, and (3) adapting to climate change. The mayors of local governments are to pledge to address these issues *en masse*, and should generally formulate a specific action plan within 1 year, issuing a monitoring report every 2 years. Participation in the CoM-Japan will be voluntary, with no duties or obligations created.

In Japan, a Feed-in Tariff system for renewable energy was initiated in 2012, and in 2014 the electric power business began to be liberalized. This made it feasible for local governments to undertake energy retailing either alone or in conjunction with other entities. For example, electric power purchases from locally produced renewable energy, which are then provided regionally, can replace electric power coming from major power producers. This results in drastic reductions in CO_2 emissions in the region. Purchasing electricity from major power producers acts as a drain, with money for electricity charges leaving the region, while purchasing electricity locally means money remains, or returns to the area. Therefore, it is anticipated that this model will be an effective means of "regional re-creation" through investments in regional renewable energy and the creation of jobs related to electric power and similar effects.

Until recently, local government climate policy in Japan has been primarily comprised of individual environmental conservation efforts and the introduction of simple technologies. Examples of this include publicity and education campaigns aimed at citizens, energy conservation and purchasing energy-efficient devices, solar power generation, and the recommended introduction of next-generation automobiles and assistance for their purchase. However, it has proven difficult to tie this as a system to large-scale reductions in CO_2 emissions. The "local energy production for local consumption" objective establishes energy autonomy, and should result in co-benefits, such as regional economic development, not just CO_2 reductions.

Moreover, in recent years, the effects of climate change have become evident in the form of rising temperatures and natural disasters resulting from extreme meteorological phenomenon. For local governments that have borne this damage first-hand and the subsequent effects, grappling with adaption measures to abate climate change is an important issue. In November 2015, the Japanese government reached

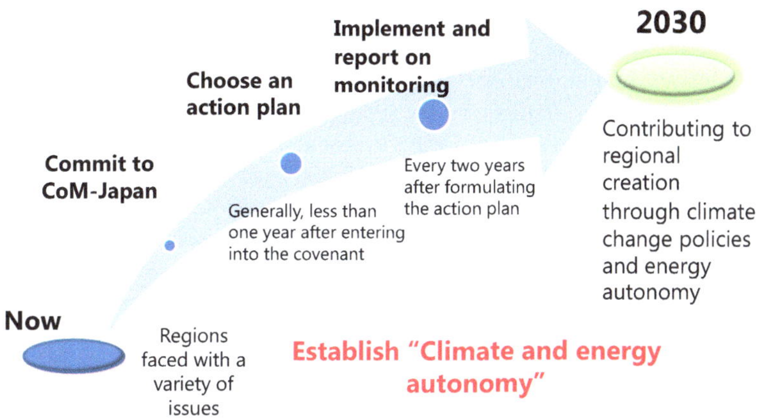

Fig. 11.1 The step-by-step process for the Covenant of Mayors-Japan

a cabinet decision, creating the "Adaption Plan for the Impacts of Climate Change" and in August 2016 announced the "Guidelines for Climate Change Adaption Plan Formulation for Local Governments", which were aimed at local governments. The Japanese adaptation law has came into force since December 2018. Because climate change risks significantly vary by region, it is critical to understand these risks, determine their priority, and tackle adaption measures. Furthermore, because the areas covered by adaption measures are wide-ranging, initiatives are needed that transcend the various areas of local government. As a result, using the CoM-Japan framework is effective for advancing policies that transcend various areas because of the leadership provided by the heads of local government (Fig. 11.1).

11.4 CoM-Japan Case Studies

11.4.1 The Five Cities of Nishimikawa Region, Aichi Prefecture

CoM-Japan was proposed at a symposium held in Nagoya in January 2015. Toyota City Mayor Toshihiko Ota, who was participating as a panelist had called that supervisors from nine cities and one village – including Toyota City – in the Nishimikawa region of Aichi Prefecture, formed a study group in March 2015. The study group convened once a month in one of the cities or village with members discussing CoM, including the relevant details and the exact aim of CoM-Japan, the capabilities of local government, the type of framework for initiatives, and other issues. The study group convened nine times through November.

Toyota City Mayor Ota said that it was easy to enter into a covenant covering only Toyota City but that this was meaningless for initiatives for the wider

Nishimikawa area. Mayor Ota suggested investigating the feasibility of wide-ranging cooperation in areas that have been effective, such as energy, transportation, and disaster prevention, then invigorating these areas in the Nishimikawa region through such initiatives. In this way, the supervisor of each local government would explain CoM-Japan to mayors of local governments and seek a decision to undertake any initiatives. Finally, in October 2015, the heads of the nine cities and one village assembled and reached a decision, with five of the ten local governments entering into the covenant. Citing large thermal power stations in their municipalities or their small size, the remaining five local governments decided that at the present stage such an initiative would be difficult and it would be risky to enter into the covenant.

On December 12, 2015, the day the Paris Agreement was adopted, the CoM-Japan covenant ceremony was held in Toyota City with the five mayors of Okazaki, Toyota, Anjo, Chiryu, and Miyoshi Cities jointly signing the first CoM-Japan. Subsequently, the five cities launched the Nishimikawa Regional CoM-Japan Promotion Conference, and in March 2017, the five cities together negotiated their SECAP.

As a "local energy production for local consumption strategy", their SECAP proposed further increasing the current 12.1% rate of introduction for renewable energy, tackling the framework construction for local energy production for local consumption. Also, because this was an opportunity to update commercial heating and cooling/hot-water supply, they were advised by experts to facilitate the introduction of highly efficient renewable energy. Moreover, as a "broad cooperation strategy" initiatives, the five cities and their 1.12 million citizens develop various efforts such as the spread of smart houses and next-generation automobiles, cool and warm shares, turning off lights, involving with citizen action groups forming networks. Plans were also developed to increase citizen familiarity with efforts to adapt to climate change. With this SECAP, the five cities adopted the goal of decreasing greenhouse gas emissions by 28% from 2013 levels over a five-year period ending in 2021. This target value would surpass the 26% decrease from 2013 levels by 2030 for the INDC (Intended Nationally Determined Contribution), which serves as the voluntary reduction target determined by Japan (Fig. 11.2).

Fig. 11.2 The Nishimikawa five cities "CoM Japan" symposium (March 12, 2017, Toyota City)

11.4.2 Takayama Village, Nagano Prefecture

With a population of 7000, Takayama Village is located in northeastern Nagano prefecture. Roughly 85% of the village is covered by forest and fields with much designated as national park lands. Agriculture is prominent in the village, especially apple and grape production, and there are also eight hot springs, so the village is both a hotspot for agriculture and tourism. In 2010, it was granted membership in "The Most Beautiful Villages in Japan",[4] and, in 2014, the entire village was registered as the Shiga Highland Biosphere Reserve. Having received these designations, in April 2016 it implemented its "Basic Environmental Ordinances". Based on these ideals, the village set goals to achieve local energy production for local consumption that utilized its plentiful natural resources. In July 2016, Mayor Katsushi Kubota entered into the second CoM-Japan.

There is a small hydropower station that utilizes an erosion control dam with a 36-m drop in Takayama Village with a maximum output 420 kW. Takayama Village provides 2% of the capital for this facility. Moreover, utilizing the Shichimi Onsen hot springs at the Shichimi Onsen Hotel, the hotel manager introduced the first binary power generation in Nagano prefecture. This is registered in the "One Village, One Natural Energy Project", implemented by Nagano prefecture. In addition to the solar panels on the rooftops of the village's primary and elementary schools, two private companies have also installed solar panels within the village. Such renewable energy generation efforts carried out in Takayama Village have created enough electricity to power 55% of the household's demand in the village.

However, presently, renewable energy generated within the village is purchased from large power companies using an FIT (Feed-in Tariff), with citizens purchasing power from large power companies and paying fees for this electricity. According to 2016 questionnaire results for the village, the charges for electricity per household were ¥167,000/year with calculations showing that 2400 of the village households paid apploximately 400 million yen annually to large power providers from outside the area. In case that a regional electricity provider established in Takayama Village, selling renewable energy within the region, would result in roughly 30% of electricity charges paid to large electricity providers being returned/remained to the region. Moreover, utilizing locally produced renewable energy could reduce CO_2 emissions from electric power originating from within the village.

Heading towards formulation of its SECAP, Takayama Village is presently holding workshops open to villager participation where participants search for mechanisms to create local energy production.

[4]A coalition aimed at "villages having magnificent local resources, however small, that are proud of their region and which seek activation through the development of community development and similar activities by citizens (NPO)."

11.5 Global Covenant of Mayors

The EU's "Covenant of Mayors", the "Compact of Mayors" that former New York City Mayor and special UN envoy for cities and climate change, Michael R. Bloomberg, advanced, and C40 (Cities Leadership Group), ICLEI (Local Governments for Sustainability), UCLG (United Cities and Local Governments), and UN-Habitat initiated in 2014 were merged in 2016, creating the "Global Covenant of Mayors for Climate and Energy" (hereinafter referred to as "GCoM"). This was the largest network of local governments worldwide for addressing climate change mitigation and adaption measures. Beginning in 2017, GCoM has been developing with the various regions not only EU but North America, Latin America/the Caribbean, Sub-Sahara Africa, India, China/Southeast Asia, and Japan in response to special institutional characteristics of each region. Preparations to launch these are being conducted in various regions globally. The CoM-Japan office which is operated by Nagoya University goes under the umbrella of the GCoM.

By participating in GCoM, studying leading cases from global networks, and developing specific local initiatives, local governments observe resilient regional development, and its promotion, resulting in high expectations for realizing low/de-carbon community development that does not rely on fossil fuels (Fig. 11.3).

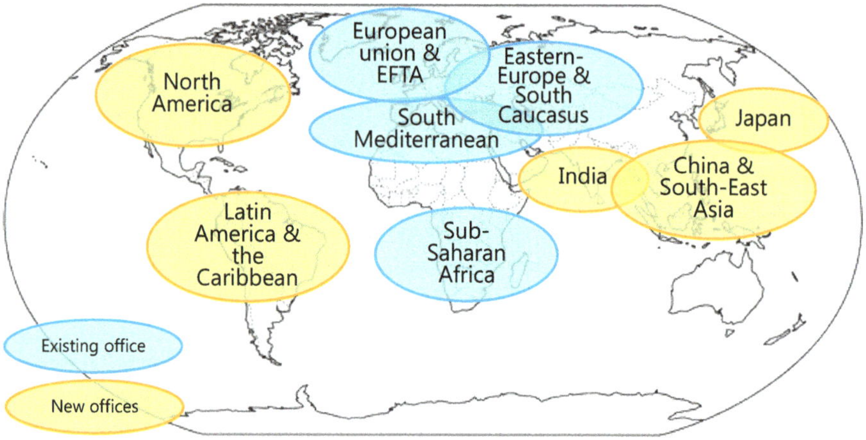

Fig. 11.3 Regional offices of Global Covenant of Mayors

11.6 Conclusion

The GCoM is the largest global coalition of cities and local governments voluntarily committed to actively combating climate change and transitioning to a low-carbon and climate-resilient region. According to the GCoM Report (2018), the Global Covenant has more than 9000 signatories in more than 130 countries across 6 continents, representing over 750 million people – more than 10% of the world's population.

As a Regional Chapter of GCoM, CoM Japan[5] has officially begun accepting and registering commitment letters from local governments from August 2018. As of the end of March 2019, there are 20 cities/towns/villages which have signed into CoM Japan (Fig. 11.4).

Since the Paris Agreement has entered into force, the role of NSAs (non-state actors) has become increasingly important. The actions of GCoM member cities will boost the achievements of each nation's INDCs (Intended Nationally Determined Contribution). In other words, the transition to a low-carbon/de-carbon society starts at the local level. This trend will further accelerate in the future, together with the achievement of the SDGs (Sustainable Development Goals) and ESG (Environment, Society and Governance) investments.

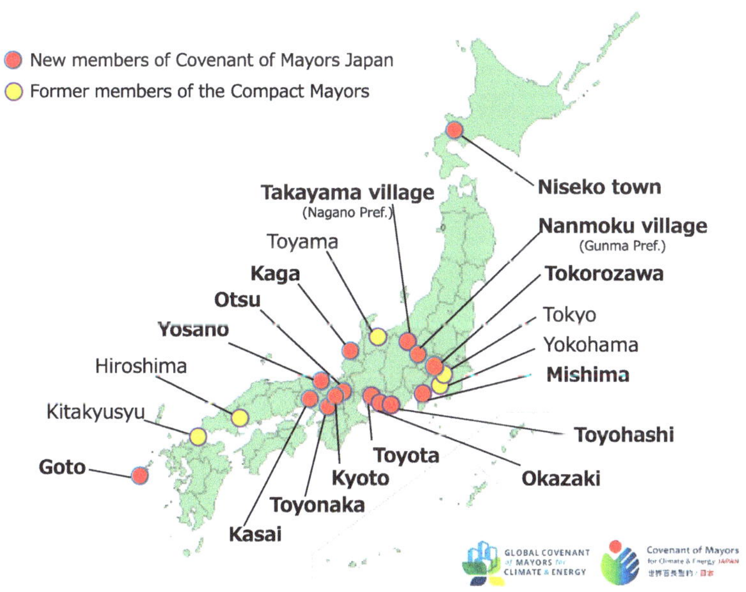

Fig. 11.4 Signatory cities of Covenant of Mayors Japan as of March 2019

[5] https://covenantofmayors-japan.jp/

Acknowledgment This research was supported by the Environment Research and Technology Development fund (1-1304) of the Ministry of the Environment, Japan.

References

Global Covenant of Mayors (2018) Global aggregation report "implementing climate ambition"

Hiroya Masuda Compilation (2014) Vanishing rural areas: overconcentration of population in Tokyo invites a sudden decrease in population. Chukoshinsho, p 256

Joint Research Centre (2016) covenant of mayors: greenhouse gas emissions achievements and projections, p 57

Ministry of the Environment support site for the creation of local government implementation plans (regional policy series). http://www.env.go.jp/policy/local_keikaku/kuiki/sakutei.html. Accessed June 2017

Sugiyama N, Takeuchi T (2008) Local policies for climate change in Japan. J Environ Dev 17(4):424–441

Chapter 12
Comprehensive Lessons Learned and the Next Steps

Mitsuru Tanaka and Kenshi Baba

Abstract As the last chapter of this book, we first summarize each chapter, then derive comprehensive lessons learned from these chapters. Summarizing the above findings, it could be said that raising resilience is a form of manifested, flexible imagination, which has led to many scientific findings and data. In order to strengthen the resilience and adaptability of regions that have signed the Paris Agreement, the above-referenced three skills must be developed to deal with the long-term, obvious risks of climate change, as well as the short-term risks hidden within regional communities. It is important to deepen the debate of the natural social characteristics of past, present, and future cities and regions, of residents' lifestyles, and of the awareness and needs of residents by understanding the external risks, vulnerabilities, and situations to be avoided proposed in this book; this can be achieved by using the related urban indicators, civil indicators, and administrative indicators adaptability tools and sharing specific, quantified evaluations, and by discussing said specific, shared methodology in scenario workshops or focus groups.

Keywords Policy transfer · The Global Covenant of Mayors · SDGs future city

12.1 Framework and Indicators of Resilience

We have developed the concept, framework, and indicators of a resilient policy model. Compared with the policy models of other organizations, such as United Nations International Strategy for Disaster Reduction (UNISDR), the United

M. Tanaka
Hosei University, Machida, Tokyo, Japan
e-mail: mtanaka@hosei.ac.jp

K. Baba (✉)
Tokyo City University, Yokohama, Kanagawa, Japan
e-mail: kbaba@tcu.ac.jp

M. Tanaka, K. Baba (eds.), *Resilient Policies in Asian Cities*,
https://doi.org/10.1007/978-981-13-8600-8_12

Nations University Institute for the Advanced Study of Sustainability (UNU-IAS), Resilient Alliance, World Bank, Arup, RPA and Siemens, Rockefeller Foundation and ARUP, Rockefeller Foundation, and Asian Cities Climate Change Resilience Network (ACCCRN), the resilient policy model has the following characteristics:

(i) We assume natural disasters and climate change, which influence each other, as two of the major anthropogenic forces (risks or stresses), impacting various facets of socioecosystems. On this premise, we define the resilient city and the scope of environmental resilience as follows: resilient city is capable of adapting to multiple environmental risks, taking into consideration the relationship between the external forces and socioecosystems; Environmental resilience should cover a broad range of issues, including energy, ecosystems, wastes, green urbanism, and others. Moreover, we assume that a combination of precautionary, adaptive, and transformative measures is required, based on the severity of the external forces.

(ii) Our primary target is local government officials who develop policies for making cities and communities resilient. Stakeholders and general citizens who collaboratively develop these policies are also included.

(iii) The policy model we proposed is a hypothetical model that describes the overall process of policy development, based on the assumption that the preparatory resilience measures and the extent of their implementation are governed by three major elements: risks of external forces, vulnerabilities, and anticipated situations to be avoided.

(iv) The policy model includes three types of indicators: urban, civil, and administrative indicators, which evaluate the state of each element: risks from natural and social external forces, vulnerabilities inherent to local communities and within the local government, anticipated situations to be avoided, preparation and implementation of resilience measures capable of addressing these.

(v) Expert knowledge, local knowledge, and living wisdom that can increase environmental resilience effectively are collected and integrated as the foundation for providing the three types of indicators. For administrative indicators, we identified 41 indicators concerning risks from natural and social external forces based on past measures, 28 indicators concerning vulnerabilities inherent to local communities and within the local government, 24 indicators concerning anticipated situations that to be avoided, and 44 indicators concerning the state of implementation and preparation of resilience measures capable of addressing these. For civil indicators, 44 indicators were re-arranged to 16 indicators by integrating similar ones to facilitate citizens' understanding. For urban indicators, over 30 indicators were initially proposed; based on the criteria of data availability, simplicity, comparability, representativeness, and balance, they were carefully validated and 18 of them were selected.

12.2 Urban Indicators

Conducting assessments using urban indicators based on public statistical information helps us understand the actual conditions and resilience of our cities and communities. Thereby it enhances our ability to resist, adapt to, and recover from devastating disasters. The outcomes of such assessments facilitate policymakers, government officers, businesses, citizens, and other stakeholders to detect weak points of the target city in comparison with other cities. In this light, various sets of indicators have been developed to assess the resilience of cities. Discussions on developing an international standard for resilience indicators have also started in a working group under the technical committee of the International Organization for Standardization. This chapter first briefly introduces the background of some important campaigns and movements for making cities more resilient. Next, two case studies of the assessment of city resilience in Japan are introduced. One is a case of introducing time-series assessments of a disaster-affected city to monitor the recovery process after a catastrophic earthquake using the CASBEE-City city-scale assessment tool. The other is a case of conducting the resilience assessment targeting major cities in Japan. These two studies show the importance of conducting the resilience assessment using urban indicators together with public statistical information.

We discussed the importance of assessing city resilience based on an appropriate set of indicators and public statistical information through two case studies conducted in Japan. While these case studies focused on cities in Japan, the methodology for developing indicators and conducting assessments on city resilience is universally applicable and relevant in any region or country. Conducting assessments using appropriate indicators and monitoring trends in city resilience will contribute to achieving resilient and safe cities.

12.3 Civil Indicators

With the civil indicators, in terms of a cognitive structure of the general public, we examine the framework of the policy model by analyzing web-based questionnaire data from nine cities in Japan. Major findings are as follows:

First, the highest risk perceived by the general public is "earthquakes," followed by "aging of population," "transportation accidents," "population decreases, low birth rates," and "infectious diseases, viruses." Perceptions of these risks vary significantly with gender, age, and region, suggesting that they accurately reflect the socioeconomic and geographical context of each region.

Second, the vulnerabilities of "high percentage of elderly population, depopulation," "high percentage of low-income earners, poor employment conditions," and "presence of earthquake zones or fault lines" are evaluated as significant.

Similar to the aforementioned risk perception, the results of vulnerability evaluations are also significantly different with age and region.

Third, "cutoff food or lifeline supply," "cutoff of energy supply," "long-term worsening of food conditions," and "long-term instability of energy supply" are evaluated as anticipated situations to be avoided. Evaluations of these situations vary significantly with the respondent's level of social capital but do not vary significantly with region. That is, they are typically shared across region.

Fourth, respondents particularly preferred "reviewing and concentration of urban functions" and "moving from high risk areas" as resilience measures by public support. Although more people preferred a self-support approach to "improving energy efficiency and promoting renewable energy" and "moving from high risk areas," for almost all other resilience measures, community support are the second most preferred after public support.

Finally, the cognitive structure whereby perceptions of risks, evaluations of vulnerabilities, and situations to be avoided determine attitudes towards resilience measures is generally supported across regions. The nature of the risk affected the object of the respondent's perceptions. For instance, disaster risks influenced assessments of situations to be avoided, large-scale risks influenced perception of topographical vulnerability, and social risks influenced perception of governmental and societal vulnerability. Topographical vulnerability was preferred to be addressed by the local community, which, conversely, may not be able to deal with crisis situations. However, the respondents may consider themselves capable of responding to social risks as individuals.

In addition to that, with the civil indicators, we also analyzed risk perception in detail. The results indicate that respondents strongly recognized the risk of disasters that frequently occur in their own residential areas as crisis events rather than other disasters. Participation in community activities and a commitment to their localities were high in rural area. The recognition of vulnerability of homes and communities was low in urban areas. These results showed that resilience index of regional community in urban areas such as Sapporo City was high in the aspect of prevention, and resilience index in rural area such as Kagoshima Prefecture was high in the aspect of cooperation and reconstruction.

12.4 Administrative Indicators

With the administrative indicators, by analyzing the questionnaire data collected from departments of planning, disaster prevention, and environment in local governments across Japan, we examine the framework of the policy model in terms of the cognitive structure of the local officials. Major findings are as follows:

First, significant differences are observed among departments for most risks, with the exceptions of "drought, depletion of water sources (water resources)," "heat waves, fierce heat," "red tides," and "harmful insects." Planning departments showed high scores on long-term situations in local communities such as population decreases and aging. Disaster-prevention departments showed high scores on not just natural external forces but also situations such as human-caused accidents and wars, and environmental departments showed high scores on situations related to typical types of pollution as well as long-term situations such as biodiversity and greenhouse gases.

Second, among all 28 indicators concerning vulnerabilities within local communities and government offices, significant differences among departments are observed on 13 indicators, such as "presence of steep terrain," "presence of rare or endangered species," and "high percentage of elderly population, depopulation," accounting for less than one-half of the number of indicators prepared. Unlike in the cases of risk perceptions discussed above, many of these indicators are common across local governments and departments, extending beyond the differences in departments' frameworks, which may result in the observed relatively fewer discrepancies among different departments.

Third, significant differences are observed among departments for most situations, with the exceptions of "long-term worsening of water resources," "worsening of living environment," and "long-term instability in energy supply." Planning departments again showed high scores on long-term situations in local communities such as "long-term decline in economic activity," "long-term decrease in the level of administrative services," "chronic worsening of public safety," and "degeneration of local culture/traditions"; disaster-prevention departments showed high scores on severe circumstances in a disaster including "direct damage to human life," and environmental policy departments showed the highest scores on "progress of global warming." These results strongly reflect the policy framework (the division of responsibilities) of each department.

Fourth, three overall tendencies can be identified for preparation and implementation of resilience measures: (i) the rates of transformative measures preparation and implementation are markedly low. Among these measures, the most frequently cited one was "development/support of next-generation energy infrastructure," followed by "centralization of urban functions (realizing a compact city)." (ii) Although all departments reported implementing similar adaptative measures, the percentage of disaster-prevention departments citing such measures was very high, while that of environmental policy departments was very low. (iii) Differences in frameworks were apparent among departments in terms of precautionary measures: 100% of disaster-prevention departments reported the "formation of disaster agreements" and nearly 100% of environmental policy departments reported that they were "promotion of energy conservation."

Fifth, when examining the correlative relationships between the number of resilience measures prepared and implemented and some factors in each department, the following tendencies can be identified: (i) As for planning departments, the factors of the risk perception and evaluation of vulnerabilities and situations to be avoided are not significant, indicating that the effects of drivers and barriers in the policy process are stronger than those of these perception factors. The drivers of "past experience/desires of residents" and "securing budgeting" have particularly strong effects. (ii) As for disaster-prevention departments, "disaster vulnerabilities" have the strongest effect and the perception of vulnerabilities increases the number of measures implemented. On the other hand, there are no apparent effects of the risk perception of external forces or situations to be avoided. The effects of drivers and barriers are also strong especially "securing budgeting" again. (iii) As for environmental policy departments, the variable with the highest value is the driver of "presence of other local governments." This tendency is distinctive to environmental policy departments, suggesting that the presence of other local governments to refer to is important in promoting policy. In many cases, policies are promoted as local governments follow each other's example. In these cases, it is likely that the presence of networks of local governments would be important as an agent to promote policy transfer. This point will be discussed in greater detail in Sect. 12.7.

As a whole, the effects of drivers and barriers in policy process are stronger to the number of measures prepared and implemented. However, even for drivers and barriers, the number of resilience measures prepared and implemented has not been fully explained. As a result, further verification of variables is needed. At this point, the results differ between civil indicators and administrative indicators.

12.5 Application of the Assessment in Japanese Cities

12.5.1 Sendai Case Study

We analyzed the results of a workshop on resilient cities held with experts and citizens of Sendai City at the 3rd UN World Conference on Disaster Risk Reduction, and collected and summarized knowledge and experiences on the extent of awareness of and learning of Sendai City's risks, vulnerabilities, situations to be avoided, and resilience measures with the status report measured by urban, civil, administrative indicators.

As a preparation, a web questionnaire survey was conducted in nine regions across Japan, including Sendai, which we have described in Sect. 12.3. The participants were recruited based on their age, gender, and disaster experience, as well as an interest in the workshop. Eighteen participants were selected. We provided the participants with four types of expert knowledge: a long-term climate projection for Sendai, the roles of infrastructure as precautionary measures, the roles of green infrastructure as adaptative and transformative measures, and the results of assessment with the three indicators. Based on this expert knowledge, the participants were divided into three groups and discussed risks, vulnerability, situations to be avoided, and resilience measures in Sendai. The main findings are as follows:

First, most participants mentioned "elevating self-support awareness and awareness of information" and "disaster preparation" as self-support measures. As for community support, while the specific suggested methods differed, most participants mentioned "creating & strengthening a sense of local community" and "local communication." For public support, most participants mentioned "providing accurate information," "using the legislative system," as well as "developing infrastructures" such as transportation, information, and green infrastructures.

Second, the participants tended to demonstrate greater expectations for public support and community support than self-support initially, but during the workshop, participants with disaster experience of the Great East Japan Earthquake indicated the importance of self-support, resulting in the attitude change of most participants. This workshop encouraged the change in participants' attitudes from public support to self-support against long term risks. It means that this type of workshop can provide opportunities to examine the adaptive risk management in a bottom-up approach and to seek a change in the manner of planning and risk management, involving both the citizens and the local governments.

12.5.2 Nagoya Case Study

Energy systems in cities and regional areas that are capable of dealing with external risks such as large-scale disasters are referred to as "energy resilience," and initiatives aimed at improving said systems are called "energy resilience measures." Energy resilience measures are largely broken into three categories: prevention, adaptation, and conversion. The prevention measures include such as the strengthening of energy supply systems, adaptation measures include such as quick recovery of damaged facilities, and conversion measures include such as conversions to decentralized energy.

In order to clarify the results of energy resilience measures, the Nagoya City Family and General Affairs Department was used as a case study. Prevention, adaptation, and conversion measures were evaluated using the following three evaluation indicators: resilience value (avoiding the costs of power outages, etc.), CO_2 reduction amount, and the amount of resiliency equipment investment. According to preliminary calculations, conversion measures were more effective than prevention measures or adaptation measures for all three categories; resilience value, CO_2 reduction amount, the amount of resiliency equipment investment, and conversion to decentralized energy were shown to be effective from an energy resilience measures. Decentralized energy conversion measures were effective not only at mitigating climate change issues, but also in terms of avoiding and abating the effects of climate change as an adaptive measure. In addition, as an example of a decentralized energy system, a model city was established to calculate the potential effects of regional energy retailers that produce local energy for local consumption. As a result, findings regarding the effects upon the regional economy and regional lifestyle changes were obtained, and this scenario was shown to have a positive economic effect.

12.6 Application of the Assessment in Asian Cities

Developing cities, generally speaking, have unmaintained public facilities that are highly vulnerable due to the expansion of densely populated areas which can be easily damaged by natural disasters. Thus, the creation of resilient cities that can minimize the risks of climate change and disasters is considered an urgent issue. Urban and regional governments have recognized present and future disaster risks, and preparations to raise urban system resilience and community resilience are necessary. Here, we looked at three Asian cities as case studies, and utilizing literature reviews, interviews, and focus group discussions by related parties, we investigated the issue of improving disaster resilience.

12.6.1 Cebu Case Study

The Philippines' Cebu City has a population of 870,000 people, and at a population growth rate of 1.9% per year, it is the second fastest growing city in the country after Manila. Urban problems include excessive land use, high migration rates, traffic congestion, and illegal squatting, and the city is vulnerable to disaster risks such as typhoons, flooding, landslides, and fires. Nationally, the foundation of systems has been strengthened on a large scale via the 2010 Philippines Disaster Risk Reduction and Management Act (DRRM) and the 2009 Climate Change Act. Using these national regulations as a template, the mayor of Cebu City established the Cebu City Disaster Risk Reduction and Management Council (CDRRMC), which is chaired by either the mayor or a designee and is organized by representatives of the Principal City Administration Department (city planning, disaster prevention, social welfare, health, police, firefighting, public utilities, public services, and financial affairs), private business owners, and citizens' groups. The council developed the Cebu City Disaster Risk Reduction Management Plan (2014–2016), which summarized policy initiatives aimed at improving disaster preparedness, response, and recovery ability, as well as the reduction of disaster risk. The construction of resilient cities involves the integration of risk evaluation and risk abatement in future city planning and infrastructure maintenance, cooperation with citizens and NGOs to implement climate-change-resistant development plans, and improvements to urban environmental problems such as poverty, energy supply, living environments, infra-structure maintenance, and waste management.

12.6.2 Nonthaburi Case Study

Thailand's Nonthaburi City has a population of 260,000 people, is considered by the government to be a center for future city growth, and is undergoing administrative, business, tourism, and residential development. This city's disaster risks include its location on low-lying plains and the presence of many natural waterways, leaving it vulnerable to frequent floods as a result of rapid urbanization, despite the fact that said waterways function as a drainage system. Floods are being prolonged, and they are causing more damage over time. The Great Flood of 2011 caused severe disruption to residents' lives and damage to property by damaging roadways and buildings, destroying crops, causing soil discharge and deterioration, and polluting the water, and many lives were lost. In Thailand, the Disaster Prevention and Mitigation Act, which established a legal framework for a national disaster management system, was enacted in 2007. In Nonthaburi City after the Great Flood of 2011, the Flood Prevention and Mitigation Center was established. The Center is headed by the mayor, and its primary duties include regulating, observing, and planning flood control measures within the city. After 2011, the city prioritized flood prevention and mitigation measures, decided upon a strategy to improve the city's resilience, and implemented said strategy. While strengthening structural countermeasures such as reinforcing river banks, constructing reservoirs, and constructing flood storage facilities, the city also strengthened nonstructural countermeasures such as developing a flood prevention decision support system, drafting a flood control management plan, implementing river basin water volume management, moving towards responsible land use, designating flood hazard areas, drawing flood hazard maps, introducing flood warning systems, and conducting flood prevention public awareness campaigns.

12.6.3 Bangkok Case Study

A recent study at the OECD has assessed disaster risk management policies to enhance urban resilience in Asian cities. The study included five city-based case analyses from multiple countries in Asia: Bandung (Indonesia), Bangkok (Thailand), Cebu (Philippines), Hai Phong (Viet Nam), and Iskandar (Malaysia). All the five cities face natural disasters risks as other Asian cities, but with substantial difference in the urban policy contexts. The study provides overall assessment and recommendations of disaster risk management policies for Southeast Asian cities, as well as individual assessment and recommendations for each case city. It highlights the need for a more comprehensive policy approach, including integrating disaster risk management into urban green growth policy frameworks. In the Bangkok Metropolitan Region (BMR), the study recommends multiple policy options, including "hard" investments in flood-resilient urban infrastructure integrated with land-use planning and zoning policies, as well as "soft" (i.e., nonstructural) instruments. In particular, the following measures for enhancing BMR's urban resilience to floods are recommended: (i) complement existing infrastructure with mangrove forests and wetlands; (ii) complement the flood control center with state-of-the-art flood simulation tools, early warning systems, and integrated data collection systems; (iii) coordinate flood-resilient building regulations in the entire BMR; (iv) develop maps of urban flood risk zones that more accurately reflect changing flooding threats; (v) promote insurance mechanisms to build greater financial resilience; (vi) use existing community entities, such as schools, religious centers, and public facilities, to increase local communities' resilience to floods and other threats more comprehensively. In this way, the successful disaster risk management outcomes will require different models, taking into account local policy contexts.

12.6.4 Baler Case Study

The Philippines' Baler City has a population of 40,000 people, and it is the capital city of the Aurora Province located in the central region of Luzon Island. Baler City is located on the coast, and with its natural splendor, it is popular for water sports tourism. Because it has wide, low-lying areas, the city is particularly vulnerable to typhoons and floods and is at risk for natural disasters. Thus, five coastal barangays (communities) were targeted for an attempt to develop a method of creating a disaster resilient community via residents' participation. This initiative used the grassroots method to facilitate knowledge and awareness on a community level, organize alongside policymakers, and participate in decision-making using the community approach in order to implement disaster risk reduction. The research contents consisted of two steps: an opinion poll of local residents regarding the reality of resilience, followed by a resident workshop to understand gaps in the current risk reduction system. The opinion poll set an index focusing on 10 processes and another index focusing on 25 outcomes; community resilience was evaluated based upon the resulting six-level score. The subsequent workshop used four factors associated with resilient communities, namely, people, community processes, organization, and resources, and evaluated each community's degree of resilience. The results of this evaluation indicated that the five communities were not capable of a strong response to disasters in terms of organization, financial affairs, infrastructure development, and institutional coordination within the municipality. This attempt to introduce interdisciplinary processes presented a usable source of information, and it clarified the mechanisms by which stakeholders can be encouraged to cooperate for the sake of disaster risk reduction. Furthermore, this study uses local knowledge and presents a methodology that takes advantage of the value of said knowledge, serving as a source of information for creating strategies for constructing communities that are powerfully resilient to disasters.

12.6.5 Findings from the Case Studies of Resilience in Asian Cities

The conclusions drawn from these case studies were that local policymakers must take up initiatives to make scientifically sound policies, demonstrate leadership, conduct efficient fundraising, facilitate cooperation between multiple centers, develop responsible land use plans, and encourage citizen participation. In enabling these important processes, the following three methods have been shown to be effective: conducting local comprehensive risk evaluations and accurately implementing the results of said evaluations when making policies, adopting a long-term approach that integrates disaster risk reduction and urban infrastructure maintenance, and strategically creating partnerships between residents and policymakers to raise resiliency.

12.7 A Platform for Policy Transfer

Various platforms are being used to disseminate the concept, policies, and tools of resilient cities to local governments around the world. The Rockefeller Group's 100 Resilient Cities is one example of this and the "Covenant of Mayors (CoM)" is an example of a climate change initiative that includes resilience. The CoM was launched in the European Union (EU) in 2008, developing into the Global Covenant of Mayors (GCoM), the largest global coalition of cities and local governments voluntarily committed to actively climate change and resilience. Local governments that participate in this framework have pledged to reduce CO_2 emissions beyond the level recommended by the EU. As a specific activity, a "Sustainable Energy Action Plan" was agreed upon in order to implement specific measures to achieve reduction goals, and a report is published every 2 years. The GCoM is a voluntary initiative in which participation is not required, and no penalties are imposed if reduction goals are not met. Challenges are expected for non-state actors, such as local governments, in the transition to low-carbon and climate-resilient cities. According to the GCoM Report (2018), the Global Covenant has more than 9000 signatories in more than 130 countries across 6 continents, representing over 750 million people – more than 10% of the world's population.

CoM Japan was established by Nagoya University in 2015 and has acted as the regional chapter of GCoM since 2017. As a Regional Chapter of GCoM, CoM Japan has officially begun accepting and registering commitment letters from local governments from August 2018. As of the end of March 2019, there are 20 cities/towns/villages which have signed into CoM Japan. Since the Paris Agreement has entered into force, the role of NSAs (non-state actors) has become increasingly important. The actions of GCoM member cities will boost the achievements of each nation's INDCs (Intended Nationally Determined Contribution). In other words, the transition to a low-carbon/de-carbon society starts at the local level. This trend will further accelerate in the future, together with the achievement of the SDGs (Sustainable Development Goals) and ESG (Environment, Society, and Governance) investments.

12.8 From Resilient City to SDGs Future City

To summarize the above findings, it could be said that raising resilience is a form of manifested, flexible imagination, which has led to many scientific findings and data. Rather than ignoring various external risks, vulnerabilities, and situations to be avoided, we should imagine every possible pattern by thinking, "if this happened this would follow," "no, this might happen instead," and "it's possible that this might also happen," and seek to investigate counter-measures for those hypothetical situations. In order to do so, the following would be required. First, we must better understand the environment around us: understanding the "structures" that support our environment, including factors such as uncertainty. Second, we must raise our awareness of and adaptability to change: signals from the future will be quicker at perceiving change, and decision makers will be able to respond quickly to change. And third, we must be aware of future uncertainties and the unknown.

It is important to understand that a "possible" future is not a prediction of what will "definitely happen." We must include possibilities in the near future and recognize potential worst-case scenarios without basing them on unlikely premises and wishful thinking. Cases at both extremes among the phenomena that are likely to occur should be considered, and if the "scope of future change" is clarified, it goes without saying that the scope of countermeasures and strategies that decision makers should think about will be clarified as well. At that time, we should no longer stick to the methods outlined in the past, but rather include a new understanding of hypothetical external forces and vulnerabilities, and situations to be avoided by adopting a view of the future, and take the first step towards creating resilient cities. This is what it means to manifest imagination.

In order to strengthen the resilience and adaptability of regions that have signed the Paris Agreement, the above-referenced three skills must be developed to deal with the long-term, obvious risks of climate change, as well as the short-term risks hidden within regional communities. It is important to deepen the debate of the natural social characteristics of past, present, and future cities and regions, of residents' lifestyles, and of the awareness and needs of residents by understanding the external risks, vulnerabilities, and situations to be avoided proposed in this book; this can be achieved by using the related urban indicators, civil indicators, and administrative indicators adaptability tools and sharing specific, quantified evaluations, and by discussing said specific, shared methodology in scenario workshops or focus groups. It is not an exaggeration to say that there is a hypothetical transdisciplinary approach for co-design and co-production by experts, policymakers, stakeholders, and the general public in this debate. Actually, various methodologies have been developed and applied to various themes (e.g. Baba et al. 2018).

In the formation of a resilient city that can deal with multiple risks, the response of local governments is determined by the risk management methods included in each government's administrative plan. How many cases are there of precautionary principles or risk management techniques being included in municipal administrative plans around the world? In Japan, environment master plans that include risk management of harmful chemicals are occasionally seen, but they are not necessarily frequent, generally speaking. Because long-term risks can influence various policy fields, precautionary principles and risk management techniques should be introduced in comprehensive plans, rather than only being included in the field of environmental policy.

An example showing this possibility can be seen in the recent Japanese policy project called "SDGs Future Cities" (The Cabinet Office 2018). This is an initiative by local governments that contribute to regional growth to accomplish SDGs aimed at mid- and long-term sustainable urban planning, and it is a model project led by the Cabinet Office. The first public announcement has already taken place, and 29 cities have been selected at 2018. Also, there are plans to announce and designate more model cities. This model project requires a vision to be drawn of the future beyond a forecast of long-term risks, but it does not present tools and methodologies to make this possible. We believe that the methodology discussed in this book may be useful in these types of projects. At the same time, these policies and tools can serve as a platform for dissemination to local governments, which could not only be effective during assemblies of model cities gathered by the government, but could also, when international NGOs like the Covenant of Mayors are formed, be effective in the establishment of an information-sharing network for governments that wish to implement comprehensive plans for future scenarios produced during transdisciplinary scenario workshops.

References

Baba K, Masuhara N, Kimura M (2018) Scenario-based approach to local water-energy-food nexus issues with experts and stakeholders. In: Endo A, Oh T (eds) The water-energy-food nexus, human-environmental security in the Asia-Pacific ring of fire. Springer, Singapore, pp 321–333

The Cabinet Office (2018) The results of selection of SDGs Future City in FY 2018. http://www.kantei.go.jp/jp/singi/tiiki/kankyo/teian/sdgs_sentei.html